...

# THE ANTI-DEATHIST

## WRITINGS OF A RADICAL LONGEVITY ACTIVIST

### Zoltan Istvan

## AUTHOR'S NOTE

While these essays have been arranged and edited for readability, many of them appear similar (if they are not new) to how they were originally published. Attempts have been made to preserve the context and moment in time they were written. Some articles contain British spelling. Publishing information and fact checks can be found by utilizing the Appendix.

# TABLE OF CONTENTS

**INTRODUCTION**

**CHAPTERS**

## III: Radical Medicine and Life Extension Science

## IV: Cryonics

## V: Existential Risk and Survival

## VI: Social Aspects of Extreme Longevity

## VII: A Politician that Wants to Live Forever

41) Why a Presidential Candidate Is Driving a Giant Coffin Called the Immortality Bus Across America

42) We Need a New Government Agency to Oversee the Search for Immortality

43): Living Forever Has Never Been More Popular

44) To Ensure a Future of Transhumanism, Atheists Should Confront the Deathist Culture Religion Has Sown

45) Now that Humans Are Living Longer, College Should be Mandatory

46) Immortality Bus Delivers Newly Created Transhumanist Bill of Rights to the US Capitol

47) How Brain Implants (and Other Technology) Could Make the Death Penalty Obsolete

48) We Must Cut the Military and Transition to a Science-Industrial Complex

49) How New Technology can Help Stop Mass Shootings


## VIII: Closing Anti-Aging & Transhumanism Ideas

50) The Longevity Peace Prize

51) Second Coming 2.0: Church Taxes Will Help Resurrect Jesus with 3D Bioprinting

52) The Future of Libertarianism Could be Radically Different

53) Singularity, Life Extension, or Transhumanism: What Word Should We Use to Discuss the Future?

54) Mind Uploading Will Replace God

*******

# INTRODUCTION

It was in English class while studying at Columbia University as an undergrad that I became a passionate anti-deathist. Our class was given a magazine essay to read on Cryonics—a procedure where newly dead people are frozen in hopes to be brought back to life sometime in the future by improved science. By the time I finished the short article, I was a converted transhumanist.

It's not that I embraced death before that moment; it was just that I didn't know that overcoming death could be a realistic life goal. But the article pointed out that not only were scientists around the world working on trying to overcome human mortality via new technologies, but that some were already doing tangible things to accomplish it.

While I support cryonics as a possible means to overcome death, it's certainly not the way I want to end up. The parts of the life extension movement that most attract me are uploading consciousness to a computer, replacing organs with bionic ones, and taking gene therapies to reverse aging.

When I first formally joined the anti-death movement in 2013 with the publication of my novel *The Transhumanist Wager*, there was a few billion dollars involved in the life extension industry. Less than a decade later, there's nearly $200 billion going towards stopping aging and overcoming death—a number sure to increase by many times over the next ten years.

There's still no guarantee people like myself won't die—in fact, it's 50/50 I still will. But I'm excited with the current developments in longevity science, and I'm just as committed as ever to the cause. Hopefully, this book of essays, which includes some of the first mainstream journalism to positively cover life extension, will help. Onward.

*Zoltan Istvan / Dec. 9, 2021*

*******

# CHAPTER I: EXPLORING THE IMMORTALITY LANDSCAPE

## 1) Death Could be a Curable Disease

Over the past few years there has been a surge in the amount of money being pumped into research on how to overcome death. Billionaires, scientists, and entrepreneurs have arrived at the revolutionary conclusion that the human body can be dramatically remade into something better, stronger, and far longer lasting.

I do believe that we will get to the stage where death will be a curable disease, thanks to technological progress. This philosophy often goes under the umbrella term transhumanism, which literally means 'beyond human.'

Transhumanism's citizen scientists, who often go by the term biohackers, promote genetic editing as a way to achieve this, turning our bodies into alien-like creatures. Others like billionaire Elon Musk think we should consider merging our brains with machines and upload our consciousness into the cloud. And some people, like Professor Stephen Westaby, want the human form to remain essentially the same but restore itself with stem cell tech and the help of bionic organs.

Whatever science transhumanists want to use to become a better species, overcoming biological death is the movement's primary goal. Most deaths in the world are caused by ageing and disease. Approximately 150,000 people die every day on planet Earth, causing devastating loss to loved ones and communities.

I think the first step in getting this figure to decrease is for governments around the world to declare ageing a disease. If society were to see ageing like it sees cancer or diabetes, then I believe more would be done to fight it.

In 2010, one of the first studies into stopping and reversing ageing in mice took place. They were partially successful and proved that 21st century science and medicine could hold the key pausing human ageing, and the illnesses associated with it. Nearly a decade later, dozens of new types of gene therapies, bionic organ experiments, and miracle anti-ageing drugs tests are underway, prompted by funding. In fact, experts including Dr. Jose Codeiro and David Wood think we're just a decade or two from significantly lengthening our lifespans.

That would be just in time too since in some countries, like the UK and USA, life expectancy has stalled after improving for decades. No one is exactly sure why but I think an increase in cheaply available junk food and obesity is one likely culprit.

Tackling ageing is just the start but eventually we'll also wipe out most diseases. To me, this is only a matter of time and there has already been research into gene editing and the impact it can have on conditions like cancer and Alzheimer's. And to rid us of the number one killer around the world – heart disease – scientists are working on bionic hearts, which are already being tested in people.

Some people worry that as we cure current diseases, new maladies or issues might appear. But, in my view, the body is essentially a biological machine and humans keep getting better every year at fixing it, regardless of what new dilemma shows up. Google Ventures' CEO Bill Maris, who helps direct investments into health and science companies, has previously said, 'If you ask me today, is it possible to live to be 500? The answer is yes.'

You might say that regardless of how amazing science is, people will always be able to die. And this is likely true. But, even in cases of acute trauma, there is great progress being made to save lives. Medicine is getting better and because of this, people are getting better at surviving. Just look to the recent past; people used to die from violence and accidents at a far higher rate than they do today. Eventually, we'll get to an era where it's very difficult to die if medical help is nearby.

One of the biggest issues in the face of all this progress is that, despite what seem like obvious benefits, conquering ageing and overcoming death are not widely supported. Around 80% of the world's population are religious, many believing in some form of life after death. Because of this, it can be argued that many see dying as part of life. This is something transhumanists refer to as 'deathist' culture.

Religion aside, some raise the argument that in eliminating death, we also eliminate value in life; these people believe death gives our lives meaning. In my mind, it's highly unlikely we'll find ourselves bored or devoid of meaning in the future, just because we don't die. Especially since in the near future — by 2050 according to Historian Yuval Noah Harari — our bodies will likely be merged with AI and we'll likely know ourselves and what we like better than we do today. It's possible by the end of the century we could be thousands of times smarter than we are now.

Besides, even if we find ways around death it doesn't mean people won't die — individuals may still want to end their lives and I'm sure they'll find a way to do so, which is why transhumanists generally support euthanasia. They also support cryothansia, where people freeze themselves while alive to be brought back in another time. It's another technology that I see increasing in popularity in the coming years.

To me, the most important part of not having to die anymore is eliminating the specter of death overhanging all our lives every minute. Humanity will be finally free of the one thing that frightens and harms us the most. There is no doubt our culture will change.

Some worry about overpopulation, something that is a drain on our planet — if no one dies, should people still be having children? Transhumanists think people will stop having kids once they know they can live hundreds or even thousands of years. People may still have children, but they might wait hundreds of years first and by then some of us may already be exploring other planets and no longer drawing upon Earth's resources.

Another institution, marriage, for example, might be transformed, as it could mean being wed for thousands of years. It's a more thorny commitment than 'death do us part.' Still, for others, the concept of the soulmate might be realized for infinity.

It's ironic that it is humans that are the thing preventing our future survival the most. Humans are a species often ingrained in their ways and beliefs. Unfortunately, when people think of humanity's future, they think of it through anthropomorphic lens—where humans always imagine themselves as mammals with the kilogram of meat we all carry around on our shoulders called a brain.

I don't think that we'll remain animals more than another half century as we discover ways via science to radically change ourselves. Therefore, we need to try to understand the future values of a world where we are literally a different being. To help, people like myself are running campaigns, workshops, and starting nonprofit entities to promote this change. We take our activism on the road to the people, and we do things like get chip implants put into us on stage (I have one in my hand).

In the end, longer lifespans will give us more control over ourselves and make it so we can spend more time with our loved ones. Extreme longevity will also add stability to economies, governments, and families. It is only by getting the world to back anti-ageing ideas and recognizing that death should be treated like a disease that humanity will be infinite.

********

## 2) Transhumanism and Our Outdated Biology

Humans are handicapped by our biology. We operate tens of thousands of years behind evolution with our inherited instincts, which means our behavior is not suited towards its current environment. Futurists like to say evolution is always late to the dinner party. We have instincts that apply to our biology in a world that existed ages ago; not a world of skyscrapers, cell phones, jet air travel, the Internet, and CRISPR gene editing technology. We must catch up to ourselves. We must evolve our thinking to adapt to where we are in the evolutionary ascent. We must force our evolution in the present day via our reasoning, inventiveness, and especially our scientific technology. In short, we must embrace transhumanism—the radical field of science that aims to turn humans into, for lack of a better word, gods.

Transhumanists believe we must stand guard against our natural genes, less they chain us to remaining as animals forever. We believe our outdated instincts can easily trick us from knowing right from wrong, practical from impractical. If one looks closely, the human body and its biology constantly highlight our many imperfections.

Compared to humans, rats have better noses for smelling. Pigeons have sharper eyes for seeing. Crocodiles can run faster. Earthworms can survive underwater longer. Cockroaches can bear far colder temperatures. Humans are only best at reasoning. Yet, computers can already beat the best of us in chess, math, and recently the sublime game Go. And the robots we've made are far stronger than we are, can handle more danger, and can fly through interstellar space without us.

Obviously, the human body is a mediocre vessel for our actual possibilities in this material universe. Our biology severely limits us. As a species we are far from finished and therefore unacceptable. The transhumanist believes we should immediately work to improve ourselves via enhancing the human body and eliminating its weak points. This means ridding ourselves of flesh and bones, and upgrading to new cybernetic tissues, alloys, and other synthetic materials, including ones that make us cyborg-like and robotic. It also means further merging the human brain with the microchip and the impending digital frontier. Biology is for beasts, not future transhumanists.

Our outdated biology's emphasis on social interaction is also dangerous for the overall evolutionary ascent of the human race—so dangerous that new questions must be asked immediately. Are so many of us healthy for this fragile planet? Should we rid ourselves of all our 25,000 nuclear weapons? Is the sexual ritual even functional anymore? Does matrimony serve purposes outside of private property and economics? Are social customs like monogamy foolishly conservative? Should we embrace a culture of drugs and biohacking? Should government use cranial implant technology to safeguard its citizens? Should society insist that all government and military leadership be equally split between females and males? Should corporations be hindered from catering to the weak, petty sides of human nature? Should religion and superstitious faiths be discouraged? These are challenging and thorny questions to ask. Yet, they should be asked, and maybe even the best answers should be implemented if we are to be true to our highest selves.

A truly transhumanist society should embrace reason and the scientific method to improve itself and bring about the best world possible on Planet Earth.

********

### 3) Silicon Valley Wants to Upgrade Pascal's Wager: New Ideas like Quantum Archaeology are Trying to Challenge Religion and Even the Permanence of Death

Since the 17th century, a core part of Western thought has been Pascal's Wager, created by French mathematician and theologian Blaise Pascal. It argues that it makes more sense to believe in God than not, since believing offers a possible way to a happy afterlife, and not believing offers either nothing after death, or maybe even eternal hell. But what happens if humans no longer age or die, something that Silicon Valley and other science hubs around the world are now spending billions of dollars trying to accomplish?

I remember learning about Pascal's Wager in Catholic grade school. As a young Christian, I became fascinated with it and its powerful logical advice. Later, as I grew out of religion and into an agnostic adult, Pascal's Wager formed the antithesis to my own personal philosophy, rendered in my book *The Transhumanist Wager*. The concept of a Transhumanist Wager argues that if it's possible to use science and technology to stop aging and overcome biological death, then that should be the primary goal and direction of one's life. This way, if people are successful in avoiding death, then they don't need to worry about what happens after it—rendering Pascal's Wager obsolete.

Right now, the ways humans are trying to overcome death are unproven. They involve using drugs and genetic editing to stop and reverse aging; cryonics where frozen dead bodies hope to be brought back to life at a later point when the technology could do it; or even data collecting of media and historical experiences of the deceased—sometimes called Mind Files—to one day recreate the person as an uploaded cyborg or AI avatar of themselves.

Regardless of the success of life extensionist's current quests to live indefinitely, the complications of Pascal's Wager in the 21st Century era don't end there. Some scientists, philosophers, and even theologians are beginning to argue for a new theoretical philosophy and science called Quantum Archaeology, sometimes referred to as technological resurrection. It's the ability to bring back the dead from any point in history, and it's trying to give a new perspective on spirituality and religion.

Quantum Archaeology has two components: the ability to reverse engineer matter and the ability to 3D bioprint that matter. Based on the trajectory of how fast 3D bioprinting is improving—a tiny human heart was printed out this year by scientists in Israel—it's hard to imagine by year 2100 we wouldn't be able to successfully print out full living human beings. Dr. Tal Dvir at Tel Aviv University's School of Molecular Cell Biology and Biotechnology, who led the team of researchers creating the bioprinted heart, believes by 2029, leading hospitals will already have organ producing printers.

The other component of Quantum Archaeology is more dubious. It relies on the universe being deterministic, which some modern physics like quantum theory has shown unlikely. However, our grasp of the universe and its rules can change as humans become more sophisticated in their research, and plenty of scientists remain on the fence about determinism. Famously, Einstein partially defended determinism, but more recently astrophysicist Neil deGrasse Tyson said it's possible we live in a simulation, which also suggests a mathematical world of causality. Also, Stanford University theoretical physicist Leonard Susskind wrote a book about how information couldn't be destroyed titled: *The Black Hole War: My Battle with Stephen Hawking to Make the World Safe for Quantum Mechanics*. Naturally, many theologians also believe in some form of determinism because they believe God is omniscient.

Putting the determinism debate aside, let's assume for a moment that humans can at least reverse engineer the past because it has definitely occurred in a specific way. This would be done by massive supercomputers untangling the subatomic occurrences of certain geographical places in the universe—in our case, planet Earth and its surroundings. It sounds insanely large and complex to do this, but already in 2018, America's largest supercomputer was able do 200,000 trillion calculations per second. Quantum Archaeology supporters argue that if the microprocessor can improve for another 100 years following Moore's Law, then the numbers of calculations a supercomputer can do could become greater than the number of atoms that compose them.

The subatomic structural blueprint of a human being in a precise moment in time is big, but not unimaginable. Mike Perry, who has a PhD in Computer Science and has studied the possibilities of technological resurrection, believes that the precise data of all humans who ever lived could fit in a nine square mile databank. The argument goes that if we had the computational power via massive data crunching to reverse engineer and record a human's subatomic structure to discover a point in time—let's say one hour before a person's death—then we could theoretically 3D bioprint those results out and that person's human life would be restored, exactly as it once was. The mind, body, and even memories would be precisely the same person, down to very molecules and atoms.

As a result of these speculative ideas, there are organizations, such as the Society for Universal Immortalism and the Turing Church— that embrace bringing back every dead person who has ever lived— all 100 billion of us since we became homo sapiens. There are also Christians transhumanists who believe in creating Quantum Archaeology so that when Jesus makes his Second Coming to Earth, this tech might be used to facilitate saving believers. Reincarnationists, such as some Hindus, are another group that are open to the idea of technological resurrection.

I'm not a believer in Quantum Archaeology. Not yet at least. Bringing back the dead is not only wildly complicated but reeks of ethical controversy. For example, it would create far worse overpopulation issues on Earth. There would be 3D bioprinted spouses meeting their loved ones who may have remarried and had new families. There would be potentiality for misuse of the technology; dictators might bioprint out clone armies of themselves and family members. And who would get resurrected? What if some people did who didn't want to be brought back to life? What would they do? And in case Quantum Archaeology isn't bizarre enough: Should we begin to put "Do not resurrect me" or "Please resurrect me" in our wills?

On the other hand, who doesn't want to bring back a loved one they lost. My father died recently from cardiac failure—he had already had four heart attacks in the two decades before—and I would love to bring him back to life. Like me, he was agnostic and didn't believe in an afterlife, but I'm certain he wanted to live longer, if his health could be improved. Naturally, that's the promising thing about the possibility of such exceptional 3D bioprinting technology. Theoretically, one's brain could be printed out identically to what it was, but one's other organs could be improved to be disease-free and rejuvenated. And it's likely by year 2100, other new life extension and anti-aging therapies would also exist to help make people younger again.

Regardless of the ethics and whether the science can even one day be worked out for Quantum Archaeology, the philosophical dilemma it presents to Pascal's Wager is glaring. If humans really could eradicate the essence of death as humans know it—including even the ability to ever permanently die—Pascal's Wager becomes unworkable. Frankly, so does my Transhumanist Wager. After all, why should I dedicate my life and energy to living indefinitely through science when by next century technology could bring me back exactly as I was—or even as an improved version of myself?

Outside of philosophical discourse, billions of dollars are pouring into the anti-aging and technology fields—much of it from Silicon Valley. Everyone from entrepreneurs like Mark Zuckerburg to nonprofits like XPRIZE to giants like Google are spending money on ways to try to end all disease and overcome death. Bank of America recently reported that they expect the extreme longevity field to be worth over $600 billion dollars by 2025.

Technology research spending for computers, microprocessors, and information technology is even bigger: $3.7 trillion dollars is estimated to be spent worldwide in 2019. This amount includes research into quantum computing, which is hoped to eventually make computers hundreds—maybe thousands or even millions—of times faster over the next 50 years.

Despite the advancements of the 21st Century, the science to overcome biological death is not close to being ready. Over 100,000 people still die a day, and in some countries like America, life expectancy has actually started going slightly backward. However, like other black swans of innovation in history—such as the internet, combustion engine, and penicillin—we shouldn't rule out that new inventions may make humans live dramatically longer and maybe even as long as they like. As our species reaches for the heavens with its growing scientific armory, Pascal's Wager is going to be challenged. It just might need an upgrade soon.

*******

## 4) A World Future Society Conference Speech: Everyone Faces a Transhumanist Wager

Recently, I had the honor to give a speech at the World Futurist Society's conference in Orlando, Florida. The World Futurist Society is the largest nonprofit organization of its kind with over 25,000 members in nearly 100 countries. Its yearly conference is a mecca for thousands of futurists looking to hear the latest forward-looking news and ideas. Hundreds of speeches, workshops, panels, meet-the-author sessions, poster presentations, and luncheons occurred.

My own speech at the conference was loosely based on an essay I recently wrote titled *Everyone Faces a Transhumanist Wager*. I wanted to share a condensed version of the talk because it presents a fundamental dilemma every human being on the planet must confront. Here's the shortened speech:

Ladies and gentlemen, we have a problem. Each one of us has a problem. In fact, no matter where you go on the planet, no matter who you find, every single person on Earth has this same dire problem.

That problem is our mortality. That problem is called death.

The reason it's a problem is because human beings love life. We all love the precious chance of existence. Even in one's darkest psychological despair, or one's most exhausting hardship, or one's most catastrophic tragedy, the thing we call life is still always miraculous. We cherish life and we don't want to lose it or have it end.

But end it will. No matter how much we wish otherwise. The stark truth is right before our eyes—that nothing in today's world can save us from death. The obviousness of this overwhelms us every time we see a loved one or a friend whose body is lifeless, never to reach out, touch, and communicate with us again. Death is final.

The great irony for our species is that we don't just have this one problem—but two problems. The second problem is nearly as vicious as the first. The second problem is the fact that most people around the world are just not worried about the first problem—they're not worried about dying. They're either religious and have the supposed afterlife all worked out, or they just don't care, or they just don't think conquering human death is possible. Whatever people's reasons, they just don't see the first problem as serious enough to warrant immediate concern—especially in a meaningful, tangible way that makes them not die. And by not recognizing death as a problem, many people have no reason to attempt to defeat it.

I have made it a mission in my life to make people aware of these two problems. It is why I wrote my philosophical novel *The Transhumanist Wager*. The concept of the Transhumanist Wager in the book is simple. It explains that in the 21st Century—the age of unprecedented technological innovation—it is a betrayal of ourselves (and the potential of our best selves) to not tackle and solve our two most pressing problems using modern science. More importantly, my book explains how we can solve these two problems.

But first, some of you are asking: What is a transhumanist? What does such a person want? What are the main goals? Some people around the world still don't know what transhumanism means. When explaining the term to people, I find it easiest to use the Latin translation. "Transhuman" literally means beyond human.

Transhumanist goals are broad and varied, but mostly they revolve around human beings using science and technology to radically improve and enhance themselves, their lives, and society. Transhumanists often concentrate on stopping or reversing aging— we are sometimes called life-extensionists or longevity advocates. Many transhumanists also focus on robotics, bionics, artificial intelligence, biohacking, and other similar fields of study. Transhumanists are often, but not always, nonreligious. They find meaning in their own lives and possibilities, without a divine creator. The philosophies of transhumanism make it possible that in the future—using extreme science and technology—one may become a so-called divine creator if they wanted. In almost all circumstances, transhumanists prefer reason over any other method of understanding to guide themselves in life.

Every transhumanist comes to their own realization of why they feel they are a transhumanist. Each path is unique, personal, and totally different than another. I want to tell you briefly about my path. I was first introduced to transhumanism as a philosophy student attending Columbia University in New York City. For a class assignment, I was told to read a magazine article on some of the recent breakthroughs in cryonics. The article described a small but passionate group of scientists who believed that science and technology would be able to bring frozen patients back to life in the future if they were preserved properly. The article also discussed the transhumanism movement, which it described as a community of reason-based futurists who wanted to use science and technology to improve their lives and live indefinitely. I was deeply intrigued. I finished that article and wanted to know more. I spent the next ten years reading everything I could find on future technologies, human enhancement, and transhumanism. I discovered the writings and work of Max More, Aubrey de Grey, Ray Kurzweil, and futurist FM-2030.

However, it wasn't until I was in the jungles of the demilitarized zone of Vietnam as a journalist for the National Geographic Channel that I came to dedicate my life to the field of transhumanism—that I came to the powerful conviction that human life should be preserved indefinitely. While in the jungle filming Vietnamese bomb diggers searching the ground for unexploded ordinances to recover and sell, I almost stepped on a partially unburied landmine. My guide pushed me out of the way, and I fell to within a foot of the mine. Tens of thousands have died from landmines in the DMZ in the last forty years, and I was lucky I was not one of them.

For me, nothing was ever the same again after that moment. The landmine incident permanently stamped into my mind how fragile the human body was—how precious our minutes alive on this planet really are. Upon returning to the Unites States, I began writing *The Transhumanist Wager.* The reason I tell you my personal story about becoming a transhumanist is that every one of us has their own story. But the two main problems we each face: death, and general apathy of death—and the choice we must make regarding them: a Transhumanist Wager—that is not just for some people. It is for every reasonable person in the world.

Indeed, in the quickly advancing 21st Century, making a Transhumanist Wager approaches us now as an ultimatum—the most challenging one we may ever face. Luckily, given how fast modern science is growing and changing our lives, making the wager is also the only reasonable option. If you love life, you will dedicate yourself to finding a way to preserve that life. Transhumanists do not want to preserve their life via heaven-promising religions, false hopes, an unconscious mystic super spirituality, or otherwise. There are only rational ways transhumanists will do it: through the tools they can create with their own hands; through the reason their brains can muster; and through the conviction their being prompts of them by not wanting to die and disappear. To do otherwise in today's world is to remain irrational and, as my novel discusses, to be masochistic and even borderline suicidal. In a world where we have the technology to travel to Mars, where we can video chat on our cell phones to someone 10,000 miles away, or we can triple the lifespan of mice with biotechnology, it's our evolutionary destiny to significantly extend our lives and to be transhuman.

Once you have identified the human race's two main problems, and you understand that you each face a Transhumanist Wager, the question is: what to do? How can you solve these problems and make the right choice in the wager.

It's quite simple, really. The journey of the transhumanist requires no ritual, no prayer, and no spiritual sacrifice or payment. It requires only your ability to reason. Ask yourself how you can best dedicate yourself to a specific cause of transhumanism and its various fields: aging research, cyborgology, stem cell science, suspended animation, singularitarianism, genetic engineering, machine intelligence, or the dozens of other areas. Then do it. For some, this may mean going into science or technology as a new career. For others it will mean volunteering in transhuman groups that need support. For some it will mean going into politics and pushing for more science-friendly laws. For others, it will mean donating resources to scientific centers and struggling innovators. For some, it will mean creating transhumanist art and using it a vehicle to push for a more scientific-minded society. For others it will mean just talking with friends and family about why you think science and technology are the best drivers of civilization.

Whatever it is that one can do, be transhumanist-minded. Be a people that belongs to a bright, rational scientific future, not one dogged by the old ways of archaic institutions, apathy, fear, or primitivism. Be transhuman, and let us all embrace our evolutionary destiny and the joys of perfect health and being that science can help us reach.

********

## 5) Forget Trump, Zoltan Istvan Wants to be the Anti-Death President

It was maddening. After departing San Francisco in September 2015 and crossing America on my campaign bus to post a Transhumanist Bill of Rights to the US Capitol, the single-page document wouldn't stick to the sandstone wall. Standing on the steps near Capitol Hill's main entrance, I began ripping off more masking tape to try to help my document adhere better. Then I heard the footsteps and yelling behind me. Posting anything on the US Capitol is illegal. Within seconds, police and soldiers carrying M-16s had me surrounded, ordering me to back away.

I turned to everyone and explained what transhumanism was: a social movement that wants to use science and technology to radically change the human species. I told them why posting the Transhumanist Bill of Rights was so important: it defended the right of humans to experiment with technology on their bodies; it gave personhood to future sapient individuals such as AI; and it established the core transhumanist aim that people have a universal right to live indefinitely through science.

A guard clutching his machine gun less than a metre away warned me I was going to be arrested. I pondered this, but turned back to the building and re-posted the document on it. The thing was, this wasn't just any document. Nor was transhumanism just any movement. Both were inescapably bound to the future of humankind. And this small act of civil disobedience was just the first step of a long journey - one of radical evolution that would involve human beings uploading their minds into machines, replacing their hearts with bionic ones and using CRISPR genome-editing tech to grow gills so they could breathe underwater. The guard looked at me as if I was insane.

In March 2013, I published a novel called *The Transhumanist Wager*. The book asks a simple question: how far would you go to fight an anti-science world in order to live indefinitely through transhumanism? Protagonist Jethro Knights would start a world war - and does so in the book. It can be seen as a political manifesto, and although I don't believe in all of the book's Nietzschean philosophy, 18 months after publishing it I announced that I was running for the US presidency. I really do want to create a science-minded world, and I think humanity's well-being and happiness would be better off for it.

Will AI solve all the world's problems when it arrives? Will sex disappear as we install microchips in our brains that stimulate pleasure zones? Will we double our children's IQ with gene-editing techniques and nanotechnology? The questions are endless, the ethics murky. Nonetheless, companies - many of which are where I live in San Francisco - are already working on all these ideas.

My goal with my transhumanist evangelism and the political Transhumanist Party I lead is to spread awareness of the questions, and - on occasion - attempt qualified answers. It's tough going, to say the least. Transhumanist activism is a new concept, and even my Transhumanist Bill of Rights won't stick to a slick historic wall.

While I never expected to win the US elections in 2016, I saw my campaign as a way to share transhumanism with the world and to help launch a crucial aspect of futurism that was missing: transhumanist activism. With two years of campaigning behind me, it's been a success, with many milestones reached.

The formation of the US Transhumanist Party in October 2014 helped launch a dozen other transhumanist initiatives around the world - including the creation of futurist parties with their own candidates. There are now a handful of transhumanist politicians running for office around the world.

Another major milestone was the Immortality Bus tour which ended in December 2015 in Washington DC. In a vehicle shaped like a huge coffin as a symbol against death, my team and I spent four months crossing America spreading the transhumanism gospel. Media attention was intense and we held rallies, staged street protests, and met the public on transhumanist issues.

In November 2015, we drove the bus uninvited to the 32,000-person strong Church of the Highlands in Alabama. The first 30 minutes went well. My team, two journalists and I wandered around the huge campus and were even given a tour by a pastor. Then the congregation members began Googling transhumanism. Within minutes the campus was put on lockdown. Gun-toting church members escorted us off the property.

Transhumanism will lead humanity forward to understand what seems like a simple truth: that the spectre of ageing and death are unwanted, and we should strive to control and eliminate them.

Today, the idea of conquering death with science is still seen as strange. So is the idea of merging with machines - one of transhumanists' most important long-term goals. But once bionic eyes are better than human eyes - something that will likely happen within the next decade or so - the elective upgrades will start. So will using robots for household chores and getting chip implants (I have one in my hand). So will CRISPR genetic editing create a new age of curing of disease and enhancing our physical form.

Embracing transhumanism will become normal, and we will become a civilisation that seeks to upgrade our bodies and lives much like we currently upgrade our smartphones.

********

## 6) To Combat Radical Violence in America, We Need Radical Medicine

America is reeling in shock from multiple shooting tragedies. The national feeling is that the violence is increasing in frequency and there's no end to the angst.

In the last few weeks, we've experienced tragic episodes in Dallas, Minnesota, Baton Rouge, and Orlando. Politicians have decried the events, calling on Congress to do more about gun control, police responsibility, and racism.

While I passionately support policies that would make Americans safer and diffuse racism, I don't think people are going to change anytime soon. However, there is something that could significantly change our national security regardless of what tragedy strikes: far better trauma medicine.

Bolstering medical treatment of extreme injuries would lessen the impact of extreme violence and shootings while the slow work of fixing the underlying systemic issues is happening.

Already today, we have the blueprints and basic know-how to create radical medicine to keep people alive under extreme circumstances. Unfortunately, as a nation, we simply are not spending the money to create it—we choose to accept a Band-Aid healthcare system that rewards medical company shareholders the more people are sick and ailing.

As my colleague at the Transhumanist Party Chris T. Armstrong said, "If you make the human body virtually indestructible, being wounded loses some of its relevance. Look at how much of the developed world now deals with AIDS. It used to be a death sentence and everyone was utterly worried about it. People are, of course, still worried about it and don't want to get it, but in America, science has mostly made it something one could live with until a cure is found."

We have revolutionary techniques like CRISPR gene editing tech, radical stem cell treatments, and 3D printed organ creation, but no one is looking at this in light of the domestic terrorism issue.

America spends approximately 20 percent of its federal budget on military and making bullets. Yet it spends only 2 percent on science and medical research. If over 10 years time, America was to divert half of those trillions of dollars for military spending into medical research, we might easily overcome death from bullet wounds, bomb shrapnel, and extreme trauma.

Already incredibly promising technology is being developed to stop death from bullet wounds. At the UPMC Presbyterian Hospital in Pittsburgh, Pennsylvania, a saline procedure where a gunshot victim's body's temperatures are severely lowered is giving doctors time to fix injuries that otherwise would prove fatal. Science has progressed so much that people who are are clinically dead can be reanimated. After their injuries are repaired, patients are slowly brought back to life.

So far, the saline technique seems to show highly positive results in pigs—90 percent of 200 pigs were able to be revived in tests. The same outcome is hoped-for with humans, but the trial is ongoing and no known cases from surgeons have been publicly reported upon. However, with luck, this potentially life-saving procedure could be rolled out across the approximately 5,000 emergency departments in America over the next decade.

Another bold new technique to help overcome extreme injury is a syringe created by RevMedx that injects tiny blood-absorbing sponges into bullet wounds. Called XSTAT 30, the syringe releases its sponge contents and can stop a gunshot wound from bleeding. This is critical, since bleeding, both internally and externally, is one of the main reasons people die from bullet wounds.

Unfortunately, gunshot victims often can't get to medical help fast enough to get some of these lifesaving technologies. In the case of the bullet wounds, drones could be used to quickly and effectively get to patients, even indoors. It's possible that flying drones could in many circumstances administer the XSTAT 30 syringe to patients too. Imagine if a hospital or police stations kept dozens of these drones on standby, ready at a moment's notice to send them out to victims. If you consider the traffic in places like downtown Dallas or New York City, drones with medicine could be the difference between life and death.

I've advocated for implants or tattoo chips on people so we could effectively know their whereabouts. It's possible to design tiny technology that resides in us that detects our pulse, so that when our pulse weakens from severe injury, it could send out signals that we are in dire trouble and need immediate assistance. The same could be done with brain implants that register extreme trauma via EEG brainwave signals.

While I'd love to stop violent terror in its tracks with better education and policing policies that make everyone get along, that is unlikely to work beyond a certain point. There will likely always be murderers, terrorists, and bad apples in society who try to ruin it for the rest of us, but we can reduce the effect of attempted killings in America with the use of transhumanist technology and radical trauma medicine by creating the medical technology that makes bullet wounds and extreme bodily injury a nonfatal issue. Unfortunately, going through any extreme physical trauma is still horrible and incredibly disruptive to life, but at least our loved ones won't be dead.

********

## 7) Why We Need a Transhumanism Movement

Recently I was asked to be a part of a debate for British think-tank Demos and their quarterly magazine. In the debate, The University of Sheffield Professor Richard Jones and I faced off over the merits and faults of transhumanism. You can read the entire debate online, but I wanted to focus on one part of it, where Jones questions why there's a need for a transhumanism movement at all.

I used to get asked the question Jones raised all the time. Luckily, the amount of people that ask it has declined, partially because transhumanism has grown so much in popularity. Many people nowadays simply accept transhumanism as part of tech and science culture.

That said, I believe it's worth sharing my thoughts on why a "transhumanism movement" is needed. Here was my answer in the debate:

I hear this question a lot: Why do we need a "transhumanism" movement? The answer, while not obvious, is straightforward. We need a movement because most have us have been brought up in cultures that don't uphold reason and science values.

In America, for example, where a majority of the population is religious, most people see no reason to want to try to use science and technology to change the human experience or overcome death. Naturally, this type of thinking carries over into the US Government, where 100% of the US Congress, the Supreme Court, and the President are religious and believe in an afterlife. This environment is not conducive for those of us who want to use science to change the world and hopefully eliminate all human ailments. It's not conducive because our leaders won't spend any money to help science and technology forward.

Government funds some medical and science research, but currently, that funding (in the US) is about 10 times "smaller" than funding for defense, wars, and bomb making. This is a tragedy that we fight wars against human beings—and not against cancer, or heart disease, or Alzheimer's, or even aging. This is a primary reason transhumanists must organize into a movement, so that we can battle the powers that be, and demand much more government funding go directly into science and medical research.

Instead of a military industrial complex, we can create a science industrial complex. If that happens, technological development will dramatically speed up—and so will all the benefits from science for the human race. We are literally in a race to save billions of lives from disease and aging.

Imagine for a moment if the environmental movement never became a formal movement. Or the women's rights movement. Organizing, collaborating, developing strategies, lobbying, and undertaking activism is how people change the world. And they do this better when they have structure and power from greater numbers of people and their groups. That's what transhumanism needs, and that's what transhumanism is essentially becoming - a bona fide global movement where political parties, television shows, nonprofits, new companies, and even wacky cross-country bus tours like my Immortality Bus occur to promote it.

Like all movements, not everything will turn out as planned. Artificial intelligence and robots may take jobs and this may cause social conflict, but transhumanism - like democracy - will find a way and improve the standard of living for everyone. Technology has that history. And it will also find a way around death and disease, especially now that companies like Google's Calico are starting to pour vast amounts of money into field.

Transhumanism is a social movement that is imbued with a techno-optimism that is contagious. The movement is growing like wildfire because people see the promise of being part of something that wants to make humans have better lives.

Movements, like transhumanism, are part of our heritage and our social evolution. People come together to make a stand and to make a difference.

About 150,000 people die every day on Planet Earth. We are like a giant organism constantly losing key parts of our bodies—parts that contain wisdom, wealth, functionality, and love. We will look back in 50 years at the phenomena of death and aging—and wonder why as a civilization we didn't do more to stop them.

Physically speaking, the human being is a machine. Like a car, it can be repaired, reshaped, and reborn. But this can only happen through science and technology. Others that claim we shouldn't worry about dying because there will be an afterlife are wishful thinking speculators. The only sure cure for death, aging, and disease can come from modern medicine and science. Everything else is faith and unproven promises.

For most transhumanists, our #1 goal is to achieve indefinite lifespans. It doesn't mean we won't want to die someday, but we want total control over such a challenging aspect of existence.

Radical science and technology will soon give us that control. It will be a bright day for those who understand that death is a tragedy—it will be a bright day for those of us who don't ever want to lose our families and loves ones.

*******

## 8) Mind Uploading: If Our Thoughts Live Forever, Do We Too?

Since there's no guarantee we will successfully cheat death by conquering aging and disease through biological experiments, we need to turn to science and technology to produce an everlasting version of the human being. Like many transhumanists, I believe we must rely on non-flesh means—think robots, AI, and other technological methods—to create a digital copy of a human that can survive forever.

With universities and tech companies building technology that could one day connect your mind directly to the internet, the debate is no longer theoretical. We need definitive answers to questions about the nature of our future digital selves, and we need them now.

Twenty-first century science has yielded many ways to replicate ourselves: We can clone ourselves (though it's illegal just about everywhere); we can implant pre-existing memories into people's brains; we should soon be able to upload the data of our consciousness to a mainframe; and engineers in Silicon Valley are already working on brain implants that allow machines to understand human thought in real time.

But is a copy of you the real you? There's a constant debate out there about whether a perfect copy of oneself—meaning an entity that contains the exact same information or data as the original—is actually oneself, or if it is just, well, a copy.

Many believe a copy is simply a secondary, inferior entity designed by a creator. Others think a copy is useful only in terms of the creator's ability to use the copy, such as growing a body and harvesting its organs for medical reasons.

Then there are people like me, who believe a copy is just as much me, as I am it.

So far, no one has successfully made a perfect copy of a human. The closest we can come is to create a clone. But since this only replicates the biology of a person, and not their thoughts and memories, a clone's traits, experiences, and memories would be different from its original, since its learned experiences and circumstances would be different.

But in 20 years, if we've perfected the technology that lets us upload our brain to a machine, that entity in the cloud might truly believe it is every bit the same as one's original self.

A perfect copy of oneself should have the same thoughts, feelings, perceptions, morals, values, and, importantly, sense of self, but that doesn't mean it has to look the same. A perfect digital copy might be a computer program designed to believe that, like its original human model, it is married with three children. A copy of a person might even exist in the form of organized subatomic particles roaming the universe via a yet-to-be-designed technology, and still believe it's human.

Yet, our memories seem the same. Our job is the same. Where we live is the same. In fact, for all practical purposes, everything is the same in the eye of the beholder. And the same goes for a digital copy of oneself. If we aren't able to intellectually explain our own consciousness, then how can we deny the consciousness of our digital copy? We have what we think and hope is free will, but it's limited to the capacity of the three pounds of meat we carry on our shoulders, which is infinitesimally small in a universe spanning many trillions of light years.

French philosopher Rene Descartes famously said, "I think, therefore I am." But the field of AI has inspired futurists like myself to argue that the real mantra should be, "I believe I think, therefore I am."

We say this because without understanding and believing that we are thinking, rational entities of ourselves, there really is no "I" that can be explained in any logical or communicable manner. Any living creature—whether human or digital—faces that same dilemma.

This is enough for me to believe that a copy of me is me—and it's enough to convince me that humans can indeed live forever. Personally, I'm not sure what type of copy I will end up pursuing, but something is better than nothing at all. Ultimately, I'd like to reach what I call the omnipotism: a post-singularity epoch where our identity, value, and intelligence control the very quarks and quantum mechanics that make up the universe. We'll barely resemble our human selves at all, but our conscious energy and thoughts will span the cosmos.

Over the next few decades, I'm pursuing a far simpler cyborg version of myself, which will be complete with bionic organs and limbs. When the tech arrives, I'll upload myself into the cloud and strive to evolve further.

But why choose only one version of myself, when you could be so many types? Maybe I'll make many copies of myself in the future. If you find a magic lamp that grants you any three wishes, game theory suggests your first wish should be for an unlimited amount of wishes. Transhumanists' central aim is survival, and many of us believe the more copies of oneself, the better.

*******

## 9) Transhumanism Is Booming and Big Business Is Noticing

I recently had the privilege of being the opening keynote speaker at the *Financial Times* Camp Alphaville 2015 conference in London. Attending were nearly 1000 people, including economists, engineers, scientists, and financiers. Amongst robots mingling with guests, panels discussing Greece's future, and Andrew Fastow describing the fall of Enron in his closing speech, event participants were given a dynamic picture of the ever changing business landscape and its effect on our lives.

One thing I noticed at the conference was the increasing interest in longevity science—the transhumanist field that aims to control and hopefully even eliminate aging in the near future. Naturally, everyone has a vested interest in some type of control over their aging and biological mortality. We are, at the core, mammals primarily interested in our health, the health of our loved ones, and the health of our species. But the feeling at the conference—and in the media these days too—was more pronounced than before.

With billionaires like Peter Thiel and Larry Ellison openly putting money into aging research, and behemoths like Google recently forming its anti-aging company Calico, there's real confidence that the human race may end up stopping death in the next few decades. There's also growing confidence that companies can make fortunes in the immortality quest.

As a transhumanist, my number one goal has always been to use science and technology to live in optimum health indefinitely. Until the last few years, this idea was seen mostly as something fringe. But now with the business community getting involved and supporting longevity science, this attitude is inevitably going to go mainstream.

I am thrilled with this. Business has always spurred new industry and quickened the rise of civilization.

However, significant challenges remain. The million dollar question is: How are we going to overcome death? It's a great question—and it's a very common question transhumanists get asked. It's usually followed by: And is it really possible to overcome death?

Honestly, no one knows the answers definitely yet, but here are the best tactics so far: Inventors like Google's Ray Kurzweil believe it can be done with machines and mind uploading. SENS Chief Scientist and Transhumanist Party Anti-aging Advisor, Dr. Aubrey de Grey, believes it can be done with biology and medicine. Others believe big data can find out the very best ways to achieve better methods for living far longer.

Organ failure is often the cause of death, and since I have heart disease running in my family, I'm a big believer in replacing organs—either with 3D printing of new organs or with robotic ones. In fact, in 10 years time, some people think it's possible the robotic heart will be equivalent to the human heart, and then people may electively seek to replace their biological heart. Because cardiovascular disease is the #1 killer in America and around the globe (claiming the lives of about a third of everyone) this type of technology can't come soon enough.

Entrepreneurs, venture capital firms, and even business media are taking notice of how new transhumanist-oriented companies are emerging and working to overcome death. The next generation of billionaires is likely to come from the biotech industry. But transhumanist technology is much larger than just biotech. It's all technology that is reinventing the human being as we know it. It's driverless cars soon to be eliminating the tens of thousands of deaths worldwide from drunk driving accidents. It's exoskeleton technology already getting wheelchair-bound people standing up and walking. It's chip implants monitoring our hydration and sugar levels, then telling our smartphones when and what we should eat and drink.

Transhumanism will soon emerge as the coolest, potentially most important industry in the world. Big business is rushing to hire engineers and scientists who can help usher in brand new health products to accommodate our changing biological selves. And, indeed, we are changing. From deafness being wiped out by cochlear implant technology, to stem cell rejuvenation of cancer-damaged organs, to enhanced designer babies created with genetics. This is no longer the future. This is here, today.

Looking forward, fortunes are going to be made by those companies that use radical science and technology to make the human being become the healthiest and strongest entity it can become.

********

# CHAPTER II: EARLY WRITINGS

## 10) Death is Not Destiny: A Glimpse into *The Transhumanist Wager*

"Death is not destiny. Death is neither inevitable nor natural," says Jethro Knights, protagonist in my philosophical thriller, *The Transhumanist Wager*.

What does Jethro mean? Death is not destiny? Death is neither inevitable nor natural?

It means, Jethro would say, that the most significant thing that has been happening to the human species is about to end.

*The Transhumanist Wager* tells the story of a man who will do anything to achieve immortality via science and technology. His main focus and drive in life is finding a way to live forever, even at the possible expense of what most people would call humanity.

When I set out to write *The Transhumanist Wager* four years ago, I did not intend it to become an edgy, controversial book. For much of my adult life, I have been a journalist covering environmental, wildlife, and human rights stories. My articles and television episodes—many for the National Geographic Channel—were welcomed in any culture and in any country. My stories were the type that a family could amicably discuss over the dinner table, or watch on television while happily cuddling together on a couch.

Perhaps it was the effect of the war zones I covered as a journalist, rising out of my subconscious, but *The Transhumanist Wager* soon took on much more contentious ideas of human endeavor and culture. For a human being, most conflict zones highlight a simple fact: Once presented with horror and death, one tends to quickly discover degrees of emotion and experience never imagined or thought possible before. For me and the difficult moments that I still vividly remember, those incidents gave me the powerful conviction that human life should be preserved indefinitely, at any cost.

Jethro Knights also realizes this early in his life, after almost stepping on a landmine in a war zone (which happened to me in Vietnam's DMZ while filming a story on bomb diggers). The revelation for Jethro is so sharp, so penetrating, so intense that nothing will ever be the same for him again.

It is from this vantage point that *The Transhumanist Wager* was written. And it is from the landmine experience that Jethro discovers the mortality crisis not only in himself, but in every human being alive. That crisis takes on the form of a wager—a choice that every human must make in the 21st century: to die eventually; or to try to live indefinitely. And if we try to live indefinitely, then we should use every tool and resource of science and technology available to us, Jethro insists. And we should do it immediately.

This is the quintessential message of *The Transhumanist Wager*. A rational and scientific-minded society owes itself the strictest dedication to applying its resources and minds to overcoming that which has been the greatest downfall of our species: our mortality.

My novel presents the story of a human being who after years of struggling, years of anguish, years of tragic loss, fights on to achieve his own immortality—and in doing so, scores a victory for all of civilization.

********

## 11) When Does Hindering Life Extension Science Become a Crime?

Every human being has both a minimum and a maximum amount of life hours left to live. If you add together the possible maximum life hours of every living person on the planet, you arrive at a special number: the optimum amount of time for our species to evolve, find happiness, and become the most that it can be. Many reasonable people feel we should attempt to achieve this maximum number of life hours for humankind. After all, very few people actually wish to prematurely die or wish for their fellow humans' premature deaths.

In a free and functioning democratic society, it's the duty of our leaders and government to implement laws and social strategies to maximize these life hours that we want to safeguard. Regardless of ideological, political, religious, or cultural beliefs, we expect our leaders and government to protect our lives and ensure the maximum length of our lifespans. Any other behavior cuts short the time human beings have left to live. Anything else becomes a crime of prematurely ending human lives. Anything else fits the common legal term we have for that type of reprehensible behavior: criminal manslaughter.

In 2001, former President George W. Bush restricted federal funding for stem cell research, one of the most promising fields of medicine in the 21st Century. Stem cells can be used to help fight disease and, therefore, can lengthen lives. Bush restricted the funding because his conservative religious beliefs—some stem cells came from aborted fetuses—conflicted with his fiduciary duty of helping millions of ailing, disease-stricken human beings. Much medical research in the United States relies heavily on government funding and the legal right to do the research. Ultimately, when a disapproving President limits public resources for a specific field of science, the research in that field slows down dramatically—even if that research would obviously lengthen and improve the lives of millions.

It's not just politicians that are prematurely ending our lives with what can be called "pro-death" policies and ideologies. In 2009, on a trip to Africa, Pope Benedict XVI told journalists that the epidemic of AIDS would be worsened by encouraging people to use condoms. More than 25 million people have died from AIDS since the first cases began being reported in the news in the early 1980s. In numerous studies, condoms have been shown to help stop the spread of HIV, the virus that causes AIDS. This makes condoms one of the simplest and most affordable life extension tools on the planet. Unfathomably, the billion-person strong Catholic Church actively supports the idea that condom usage is sinful, despite the fact that such a malicious policy has helped sicken and kill a staggering amount of innocent people.

Hank Pellissier, a futurist and organizer of the conference Transhuman Visions, says, "The public majority disapproves of Christian Scientist and Jehovah's Witness parents who deny medicine to children afflicted with life-threatening illness. The public regards the anti-science attitudes of these faiths as unacceptable. Likewise, we should similarly disapprove of the withholding of any medicine or life extension practices that deter death for individuals, of any age."

Regrettably, in 2014, America continues to be permeated with an anti-life extension culture. Genetic engineering experiments in humans often have to pass numerous red-tape-laden government regulatory bodies in order to conduct any tests at all, especially at publicly funded universities and research centers. Additionally, many states still ban human reproductive cloning, which could one day play a critical part in extending human life. The current US administration is also culpable. The White House is simply not doing enough to extend American lifespans. The US Government spends just 2% of the national budget on science and medical research, while their defense budget is over 20%, according to a 2011 US Office of Management Budget chart. Does President Obama not care about this fact, or is he unaware that not actively funding and supporting life extension research indeed shortens lives?

In my philosophical novel *The Transhumanist Wager*, there is a scene which takes place outside of a California courthouse where transhumanist activists are holding up a banner. The words inscribed on the banner sum up key data:

*By not actively funding life extension research, the amount of life hours the United States Government is stealing from its citizens is thousands of times more than all the American life hours lost in the Twin Towers tragedy, the AIDS epidemic, and the Vietnam War combined. Demand that your government federally fund transhuman research, nullify anti-science laws, and promote a life extension culture. The average human body can be made to live healthily and productively beyond age 150.*

Some longevity experts think that with a small amount of funding—$50 billion dollars—targeted specifically towards life extension research and ending human mortality, average human lifespans could be increased by 25-50 years in about a decade's time. The world's net worth is over $200 trillion dollars, so the species can easily spare a fraction of its wealth to gain some of the most valuable commodities humans have: health and time.

Unfortunately, our species has already lost a massive amount of life hours; billions of lives have been unnecessarily cut short in the last 50 years because of widespread anti-science attitudes and policies. Even in the modern 21st Century, our evolutionary development continues to be significantly hampered by world leaders and governments who believe in non-empirical, faith-driven religious doctrines—most of which require the worship of deities whose teachings totally negate the need for radical life extension science. Virtually every major leader on the planet believes their "God" will give them an afterlife in a heavenly paradise, so living longer on planet Earth is just not that important.

Back in the real world, 150,000 people died yesterday. Another 150,000 will cease to exist today, and the same amount will disappear tomorrow. A good way to reverse this widespread deathist attitude should start with investigative government and non-government commissions examining whether public fiduciary duty requires acting in the best interest of people's health and longevity. Furthermore, investigative commissions should be set up to examine whether former and current top politicians and religious leaders are guilty of shortening people's lives for their own selfish beliefs and ideologies. Organizations and other global leaders that have done the same should be scrutinized and investigated too. And if fault or crimes against humanity are found, justice should be administered. After all, it's possible that the Catholic Church's stance on condoms will be responsible for more deaths in Africa than the Holocaust was responsible for in Europe. Over one million AIDS victims died in Africa last year alone. Catholicism is growing quickly in Africa, and there will soon be nearly 200 million Catholics on the continent.

As a civilization of advanced beings who desire to live longer, better, and more successfully, it is our responsibility to put government, religious institutions, big business, and other entities that endorse pro-death policies on notice. Society should stand ready to prosecute anyone that deliberately promotes agendas and actions that prematurely end people's useful lives. Stifling or hindering life extension science, education, and practices needs to be recognized as a legitimate crime.

*******

## 12) Should Transhumanists Have Children?

Transhumanists are people who desire to use science and technology to improve the human being. While the international movement of transhumanism is rapidly growing and diversifying, its most important goal remains the same: overcoming human mortality. Many experts believe some sort of indefinite sentience for individual human beings, whether via age reversal or by mind uploading into computers, will be achieved around 2045. Such incredible advances will change the way the species views itself. Procreation, the foundation of human civilization, will be one activity that is dramatically affected.

If all goes well, my wife and I will be bringing a baby girl into the world in a few days time. It will be our second child. I'm often asked whether it makes sense for transhumanists to have children. For a group of people who mostly doubt they will ever die, it's a valid question. Doing away with death presents a historical quandary for the human race. For example, if you were to live indefinitely, would you have children within the first 250 years of your life? However, since many people probably wouldn't be biological in 250 years time — people that far in the future will likely have become all-digital by uploading themselves into machines — the question itself is turned upside down. Such is the nature of the budding movement of transhumanism, which is on course to create a paradigm shift for civilization in the next half century.

As a transhumanist, I choose to have offspring for many of the same reasons other people do. I want children because they're amazing, thrilling, and beautiful to nurture and raise to their best potential. It completes the artist in me like no pen and paper ever could. Moreover, the love shared with one's child is very comforting and precious. It bonds my family together, giving everyone involved — including my parents — profound meaning. Of course, being a father is incredibly fun, too.

Having a child is also one way to achieve a sort of immortality; if I was to die, at least my genes (and hopefully some of my ideas) would be carried on. As a transhumanist, I don't consider that an acceptable form of immortality, but I find some consolation in it, anyway.

The question of whether transhumanists should have children is ultimately a personal and subjective one. In the near future, concepts of procreation will drastically change. Already, advanced in vitro fertilization techniques, prenatal testing, and genetic engineering are altering the way we approach the procreation process. In the next two decades, advances in cloning and ectogenesis — using an artificial environment to grow life outside of where it would normally be found — will further our perspectives and present many philosophical and moral challenges. Yet, all this innovative procreation science still remains within the confines of our biological human parameters. What happens when we leave our genes and DNA for a virtual existence in machines?

"Mind uploading may be here in 30 years or so," says Avinash Singh, a software engineer and co-founder of the India Future Society. "Once inside a computer, we may have brought the essence of our biology with us, but it won't remain with us for long. We will quickly adapt and evolve in the virtual world, especially with the help of increasingly powerful artificial intelligences."

If we transcend our biology completely, does this mean we won't have incentive to procreate? Will human beings living exclusively in computers really drop certain rituals that stem from millions of years of evolution? The likely answer is yes. Over time, we'll probably program the desire for progeny out of ourselves. Procreation in the sense we know it — along with sex — will likely become obsolete. Indeed, even the concepts of male and female will probably disappear unless a reasonable purpose inside the digital frontier is found for either.

Leaving behind our biological propensities and heritage has nothing to do with right or wrong, but rather whether it's useful in a digital environment given our current evolutionary incentives. Such a computational-run world may at first seem alien, shallow, and devoid of compassion, but the nature of our digital selves will not know that. We'll be far from human by that point.

We can anthropomorphize our future digital selves all we want in hopes we'll be able to maintain our humanity, but once permanently in a machine, our mammalian proclivity will quickly become as foreign to us as an infant's perspective is to an elder's perspective. After that, it won't take long before evolution makes us virtually unrecognizable to our former selves. Our digital avatars will adapt and advance at evolutionary speeds never known before.

Digital environments will likely become the playgrounds of personal egos and their wills, where self-centered domination of perspective and experience are paramount, as detailed in my philosophy TEF, which stands for Teleological Egocentric Functionalism. TEF was designed as a bridge from today's Homo sapien to tomorrow's digital avatar who wields the power of extraordinary machine intelligence. From there, it's but a short leap to the Singularity.

As a transhumanist, I look forward to the coming future and all its grand possibilities, including my existence in a machine. In the meantime, I'll be grateful to welcome my daughter with much love and care into our biological world.

*******

## 13) Don't Want to Die?
## Support a 1 Percent Jethro Knights Life Extension Tax

Most people don't read scientific journals, but if they did they'd know the transhumanist field of life extension is repeatedly making major scientific and medical breakthroughs. Life extension science — also called longevity research, biomedical gerontology or anti-aging medicine — is the study of keeping people alive. It especially focuses on slowing down or reversing the aging process in order to extend lifespans. Few fields of study offer so much for civilization. Some results of life extension science are already being realized, and lives around the globe are being lengthened, better health is being achieved, and disease is being controlled or eliminated.

It all sounds great. Unfortunately, it's not. In our modern world of smartphones, robotics and jet travel, there are still 150,000 deaths a day. Furthermore, poor health and disease afflict billions of people, causing many of them to suffer. In the next 10 years, over a half-billion people will die, many enduring horrifically painful and degrading deaths as their bodies and brains stop functioning. It doesn't need to be that way.

If human society values the sanctity of life, it must ask: Is there another way? The answer is surprisingly simple: Yes, there is. Support a one-time 1 percent Jethro Knights Life Extension tax applicable to every adult human being on the planet. Dedicate just 1 percent of your personal net worth for life extension science, and in 10 years time, medicine, longevity and human health will be transformed. The world can conquer death in about a decade's time if enough resources are put towards it.

Some of you are asking: Who is Jethro Knights? Jethro Knights is the philosopher protagonist in the novel *The Transhumanist Wager*. What makes him important is he symbolizes every person on the planet, each who faces the same universal dilemma: human mortality. No sane and reasonable person wants to die if it can be avoided. In the 21st century, the age of unparalleled scientific and technological achievement, many scientists and futurists realize we are very close to medically overcoming biological death. Our human bodies and minds can be made to live indefinitely. Jethro Knights is the person who declares death can be scientifically overcome, and then makes an oath to do everything in his power to achieve this.

Few people want to address the fact that science and medicine are lagging far behind where they could be if adequate resources were given to them. Even fewer people would agree that they are responsible for that fact. But make no mistake: We are all responsible. To some extent, we are all responsible for our own deaths and suffering. We have not dedicated enough of our time, energy and resources to the advancement of science and medicine. We support trillion dollar wars in far-off lands, but not trillion dollar wars at home against cancer, heart disease, diabetes or aging.

There are many excellent research groups, university departments, and companies around the world trying to eliminate aging and disease. SENS Research Foundation, Google's Calico, Buck Institute for Research on Aging, Maximum Life Foundation, Human Longevity, Inc., and USC Davis School of Gerontology are some of them. But all of them need more resources to speedily tackle the complexity of human longevity. Yet, nobody is giving enough money to these scientists and entities because nobody cares enough. The U.S. Government isn't doing much either to extend American lifespans. They spend just 2 percent of the national budget on science and medical research, while their defense budget is 20 percent, according to a 2013 Center on Budget and Policy Priorities chart.

What people don't realize is that with enough research money properly focused — $50 billion dollars, some experts say — human aging and the terror of disease can likely be halted. With a trillion dollars (the world's wealth is over $200 trillion), human death can be halted in 10 years time.

In the end of the day, controlling aging and disease are just more science puzzles waiting for the modern world to solve. We could help that process along if we changed the psychology of civilization's culture — a culture that largely believes human death is unstoppable and inevitable. Aging should be seen as a fixable problem, not as a destiny. The human race can overcome its biggest natural hurdle.

For all these reasons, I propose the 1 percent Jethro Knights Life Extension tax. Nobody likes taxes, especially not new ones. However, some taxes serve their purpose and can also change the world. Naysayers will say: Not everyone will support such a tax. That is certainly true. Perhaps a general vote could decide this. Others will say: Even if there was such a tax, many people will not pay it. That is also true. Nonetheless, only a fraction of this 1 percent world tax would need to be used to eliminate disease and aging in a decade's time.

Many other pertinent questions exist, too. Which scientists, universities, companies and longevity research groups would get the money? How would governments coordinate such a massive international endeavor? What life extension strategies would be most funded? Who or which countries would get the longevity benefits first? And how could the planet handle more overpopulation? These are questions to be answered by politicians, philosophers, scientists and the people of the world. But, rest assured, they all can be satisfactorily answered.

Most likely, an intergovernmental body, perhaps similar to the World Health Organization, would need to be set up to best coordinate and achieve such radical life extension aims. Of course, the first step to such a noble goal for the human race starts with a personal decision on the part of every human being. If you love your life, you will want to live longer and contribute towards conquering human mortality. You will want to support a one-time 1 percent life extension tax to end disease and death for yourself and your loved ones.

********

# 14) A TEDx Talk on Life Extension

I recently had the opportunity to be the closing speaker at the 5th annual TEDxTransmedia event, held in the iconic Radio Television Suisee building in Geneva, Switzerland. Organized by media pioneer Nicoletta Iacobacci, the event was opened by a two-foot tall robot that gave a short welcome speech. The theme of the event was exponential beauty, and over a dozen speakers, performers, and young change makers also made presentations. The event was an overwhelming success that was topped off by a festive farewell cocktail reception.

Below is a condensed version of the first part of my speech, which touches on various ideas of beauty and is dedicated to the field of life extension science. My talk was titled: The Beauty of Being Alive:

We've all heard the famous saying that "beauty is in the eye of the beholder." I tend to agree with those words. However, what often goes unmentioned is another truth about beauty: In order to experience beauty, we must be alive. We must have blood pumping through our veins. We must have air filling our lungs. We must have a working brain that is conscious of existence. All this is necessary to experience beauty. Without it, beauty would not exist for us.

In the last few hours you have heard many incredible speeches about beauty. You have heard many definitions, many interpretations, and many ideas about it. The experiences we have of beauty seem endless. For some, it's a spectacular sunrise over the snow covered Alps. For others beauty is simply being kind to strangers. For some, beauty is racing a sports car down an empty country road.

I want to tell you about a unique version of beauty I know. I want to tell you about the most beautiful moment I've ever had—a moment when I almost lost my life.

For a number of years, I worked regularly as a journalist for the National Geographic Channel. Some of my reporting put me in danger, such as when I tracked illegal loggers in Paraguay, volcano boarded in Vanuatu, or chased wildlife poachers in Cambodia. Some of my assignments were even in war zones, like the ongoing conflict in Pakistan and Indian Kashmir.

However, it was in the jungles of Vietnam that I faced my most dangerous assignment. During the Vietnam War, America dropped millions of bombs on Vietnam. Approximately 5 to 15 percent of those bombs never exploded. Many farmers in Vietnam grow rice and earn only a few dollars a day. Yet the salvaged metal from one large unexploded bomb can equal almost a year's worth of rice farming. Not surprisingly, thousands of farmers have given up their agriculture ways to search for unexploded bombs in the jungles, creating the strange profession of bomb hunters.

The problem with this line of work is twofold. Parts of the country are filled with landmines, and dismantling armed bombs can be extremely dangerous. Despite this, bomb hunting in Vietnam is thriving. And fortunes have been made.

I went to Vietnam to document this fascinating story. After almost a week of roaming around the countryside with a group of fearless bomb hunters in the DMZ (demilitarized zone), I was mentally exhausted. There are few things worse than walking through the jungle, wondering if the next step you take is going to be your last because you got unlucky and triggered a mine.

On my final day of filming, we ventured back into the jungle hillsides. I left the trail by just a few steps to search out a strange impression in the ground. It looked like a large object may have hit the earth in that exact area decades back. Without warning, from behind me, my interpreter tackled me. My video camera and I slammed into the ground. Everyone began shouting and pointing. In the earth, right in front of where I was walking was a small round buried object. It was a partially exposed landmine.

After being partially tackled, I lied still on the ground, staring at the landmine, which was only inches from me. While the mine did not go off, something in my head did. For me, nothing was ever the same again after that moment. The incident permanently seared itself into my psyche, reminding me of how precious our minutes alive on this planet really are. What happened to me can be aptly described as a philosophical bomb—which is why I consider it the most beautiful and transformative moment of my life. It triggered a response that I am still passionately pursuing today: the goal of eliminating human death and suffering.

Upon returning to the United States, I curtailed much of my journalism work, and began actively promoting the field of life extension science and transhumanism, areas of research that aim to use radical science and technology to stay alive as long as possible and improve the human condition. Many top scientists believe we have a good chance at conquering human death via modern medicine within 20 years. With enough resources directed at the research, we could probably do it in 10 years.

What is happening to our species is we are becoming far more than just human. All around us incredible advances in technology are changing and improving us. The advocates of this type of human progress are scientists, technologists, futurists, medical professionals, engineers, and everyone interested in using science and technology to significantly advance and evolve the human being. When you get a 3D printed organ to save the life of your child—that is it. When you turn on your cell phone and call someone on the other side of the planet—that is it. When you can't see because your vision has deteriorated, and a surgeon fixes your eyes with lasers—that is it. This futuristic and radical technology is all around us, and it's advancing exponentially.

A new era is upon us; an era where we suffer far less from the hardships our ancestors faced. Look anywhere around you: Paralyzed people in wheelchairs can walk via exoskeleton technology. Those born deaf can hear via cochlear implants. Those with memory loss from Alzheimer's can remember with brain chip implants. Those who lost arms in war zones and automobile accidents have robotic limbs with feeling sensors that are tied to their nervous systems. Indeed, modern science and technology are slowly eliminating all suffering and disease on our planet.

Despite the fact that the world is beginning to embrace such amazing innovation, some people argue that death is both necessary and beautiful. Some people feel death fits naturally in the cycle of human life. Others insist there would be no value or meaning in living without death. I disagree. In fact, I strongly disagree. Death is not beautiful at all. Perhaps the way a person dies can be beautiful. But the result afterwards, of not speaking to them again, of never seeing them again, of never holding them again—this is not beautiful.

Additionally, the idea that death is necessary to make life meaningful is also incorrect. Our lives have plenty of meaning and value without the spectre of death hanging over us. Consider this: Except for mean-spirited or insane people, few would endorse their loves ones (such as their children) dying one day just to justify the so-called meaning in their loved one's lives. Having meaning in one's life is not defined by whether one dies or not, but by whether that person chooses to see and find purpose in their daily existence, whatever that may be.

Billions of people unknowingly live in an indoctrinated global culture of death. Often, this is promoted by religious ideas which endorse afterlives. But cutting-edge life extension science is not interested in afterlives it can't see, know, or prove, but rather in what is taking place today: the living.

As a nonreligious person, I believe beauty (and meaning) end with death for any person who has once experienced or pondered it. This all means one very important thing: Preserving human life, memories, consciousness, and the general essence of a person is an imperative for beauty to exist to that beholder. Life extension science and technology are obvious ways to do that, and should be actively supported and pursued by society. Finally, death—especially in the 21st Century—should be seen for what it is: the ugliest thing we know in the universe.

*******

## 15) Should a Transhumanist Run for US President?

I'm in the very early stages of preparing a campaign to try to run in the 2016 election for US President. I'll be doing it as a transhumanist for the Transhumanist Party, a political organization I recently founded that seeks to use science and technology to radically improve the human being and the society we live in.

In addition to upholding American values, prosperity, and security, the three primary goals of my political agenda are as follows:

1) Attempt to do everything possible to make it so this country's amazing scientists and technologists have resources to overcome human death and aging within 15-20 years—a goal an increasing number of leading scientists think is reachable.

2) Create a cultural mindset in America that embracing and producing radical technology and science is in the best interest of our nation and species.

3) Create national and global safeguards and programs that protect people against abusive technology and other possible planetary perils we might face as we transition into the transhumanist era.

These three goals are so simple and obvious, you'd think every politician in the 21st Century would be publicly and passionately pursuing them. But they're not. They're more interested in landing your votes, in making you slave away at low-paying jobs, in keeping you addicted to shopping for Chinese-made trinkets, in forcing you to accept bandage medicine and its death culture, and in getting you to pay as much tax as possible for far-off wars (places where most of us will never step foot in).

While I look forward to the challenge of being involved in politics, the reality is, of course, that it's totally improbable a new independent party and its candidate will get elected. It probably will be impossible to even get on many state ballots. Obviously, I'm aware of that. Why do it then? Because it's a start. And if transhumanists—a growing group consisting of futurists, life extensionists, biohackers, technologists, singularitarians, cryonicists, techno-optimists, and many other scientific-minded people—are serious about the pending future, then it's time to get involved in the political game. Enough lawyers and politicians over the years have had their fill. Enough faith-touting Congress members have stifled this country. Enough WASP-supporting men have sat and tried to dictate its terms for too long. Transhumanists must get involved to protect and usher in their futures.

The transhumanism movement goes back decades, to a time when philosophers, futurists, and scientists began understanding how fast technology could solve all the world's problems. The movement has continued to grow and is now spreading amongst the youth like wildfire. There are many employees at major tech companies like Apple, Facebook, and Google who subscribe to transhumanist aims. Transhumanist-themed conferences, groups, and even schools, like Singularity University, are popping up. Notably, even Italy has recently elected a transhumanist Parliament member named Giuseppe Vatinno. Another transhumanist politician Gabriele Rothblatt is running for US Congress in Florida's 8th District.

The term transhumanism is, of course, just a label, like Republican, Democrat, libertarian, or tea partier. But the label is important, nonetheless. This is because there is a big difference between those who appreciate science, and those who are willing to use it to fundamentally alter themselves and society in the 21st Century. In the next 25 years, all of us will face a choice about how far we want to take technology and science—all of us will face a Transhumanist Wager. Artificial hearts will become better than the best human hearts. Bionic arms will become superior to human arms. Smart phones will become the size of a fingernail and will likely be implanted into your body. Speaking out loud will disappear as the modern world uses mindreading headsets to communicate, which already exist. Where will you stand? How far will you take technology in your life?

The future is less about social security, climate change, immigrant border traffic, taxes, terrorism, the economy, and the myriad other issues that flash across news headlines every day—and more about how far we are willing to use science and technology to fundamentally alter the human being and experience. That transhuman future doesn't care what color skin you have, what your sexual orientation is, what faith you embrace, where you were born, what type of job you have, or what political perspective you hold.

Transhumanism is not a political end, but a life-affirming commitment to becoming one's best self using the help of reason, science, and technology. It doesn't belong to any formerly established political system or party. It's only concerned with reaching its goals and being true to its ideals. My own politics are similar; I'm interested in making the most headway we can into our future.

Friends, we've all seen what's happening lately. The transhumanist era is literally upon us. Those paralyzed and bound to wheelchairs can walk via exoskeletons. Those who have never heard sound can now hear via implants. Gun shot victims who are dead are brought back to life via suspended animation. The poverty rate is the lowest it's ever been around the world. Science and technology are responsible for these joys and successes.

Yet, we spend so little of our resources on the brilliance that science can bring all of us. America still spends almost 10 times its resources on defense than on science and medical research. It spends approximately four times its resources on the prison system than on education for our kids. It spends at least 100 times its wealth on bureaucratic-inspired legal fees than on critical life extension science to keep its citizens alive. This is why a transhumanist politician could be an extraordinary help to the nation and the 21st Century.

The future is coming so fast, that unless we change the paradigm and how we lead our great nation forward, we will find ourselves at the mercy of such powerful technology. It's time to introduce the future to our political leaders who will lead this country forward. The Transhumanist Party will not win this election. But it can change the questions the real elected leaders will ask. That is something significant, indeed.

*******

## 16) Origami Cranes: Who is Responsible for this Child's Death? (Introduction to the World's First Mainstream Media Column on Transhumanism: *Psychology Today's: The Transhumanist Philosopher*)

Colorful origami paper cranes appeared on a neighbor's front yard last week, as they often do on lawns across America when a child is dying from a brain tumor. The cranes are supposed to be a heartwarming symbol of eternity, life, and good luck, put up by family and friends to support that child.

Today that child died.

I live in a community of tree-lined streets in San Francisco where many kids go to college and pursue careers in technology, law, and medicine. It's a close-knit neighborhood with few issues. A sick child here usually gets the best healthcare possible. Two of the world's leading medical centers, UCSF and Stanford, are within close driving distance.

Unfortunately, when illness struck this 6-year-old child, the best medicine was not enough. Some people find it hard to believe in our modern world of smart phones and jet travel that we still can't stave off disease. Inevitably, everyone asks: Who or what is responsible for the death of this child? The answer is simple: *We* are all responsible.

Few people want to address the fact that science and medicine are lagging far behind where they could be if adequate resources were given to them. Even fewer people would agree that they are responsible for that fact. But make no mistake: We are all responsible. We are all responsible for the death of that child. We have not dedicated enough of our time, energy, and resources to the advancement of science and medicine. Furthermore, every time we give a dollar to a religious institution instead of to a scientific institution, every time we endorse a politician who cares more about lobbyists than our fumbling national education system, and every time we support our government's trillion dollar wars instead of a trillion dollar war on cancer, heart disease and diabetes, we support the premature death of innocent people.

I recently took a tour through the research center at SENS (Strategies for Engineered Negligible Senescence), one of the most prominent nonprofit scientific organizations attempting to stop aging and disease. Filled with white-gowned men and women bent over microscopes, SENS has many promising scientists in its labs. I asked Dr. Aubrey de Grey, Chief Science Officer of SENS, and one of the most visible anti-aging advocates in the world, what the SENS budget was for 2013.

"Less than five million dollars," he told me.

I gasped. That's tiny, I thought. Many nonprofit organizations like World Wildlife Fund, Feed the Children, and Red Cross have at least twenty-five times that amount for their annual budgets. And those organizations are not trying to stop the human race's most significant problem: dying.

There are many excellent research groups around the world trying to eliminate aging and disease, but all of them need more resources to speedily tackle the complexity of human longevity. Yet, nobody is giving enough money to these scientists because nobody cares enough. The US Government isn't doing much either to extend American lifespans. They spend just 2% of the national budget on science and medical research, while their defense budget is over 20%, according to a 2011 US Office of Management Budget chart.

What people don't realize is that with enough research money properly focused—$50 billion dollars, some experts say (the world's wealth is over $200 trillion)—human aging and the terror of disease can likely be halted. In the end of the day, controlling aging and disease are just more science puzzles waiting for the modern world to solve. We could help that process along if we changed the psychology of civilization's culture—a culture that largely believes human death is unstoppable and inevitable. Aging should be seen as a fixable problem, not as a destiny. The human race can overcome its biggest natural hurdle.

My new *Psychology Today* blog is titled *The Transhumanist Philosopher.* Every few weeks, I will be writing about how individuals and society are being transformed through rapidly advancing science and technology. I will be bringing you stories that dive into philosophical, sociological, and psychological perspectives of human enhancement, longevity issues, and transhumanism. I will be interviewing futurist and science leaders. I will be reviewing their books and projects. I will be exploring the philosophy of ending human aging and embracing indefinite lifespans.

One theme in my blog will always remain prominent. If, as a society, we choose to begin spending our energy and resources on health and longevity, then we will soon achieve the promise that life extension research, transhumanism, and human enhancement can bring us. We will soon become all that the human being is capable of becoming.

My goal of this blog is to quicken the coming of the day when there will be no more origami paper cranes on anyone's lawns ever again.

*******

# CHAPTER III: RADICAL MEDICINE AND LIFE EXTENSION SCIENCE

## 17) Which New Technology Will Win the Race to Repair and Replace our Organs?

An extraordinary competition is underway—one that could be more impactful to the human species than any other technological rivalry to come before it. Soon, the radical concept of substantially improving or outright replacing our organs is going to be commonplace.

Globally, organ failure is a leading cause of death. But transplantable organs are in far too short of a supply around the world to help many in need—even former Vice President Dick Cheney had to wait 20 months to get his new heart.

Various methodologies, technologies, and even spiritual and philosophical preferences are dividing up this human upgrade quest. Companies are launching into the field, hoping to create the dominant longevity tech that people in the near future will use to live to 150 and beyond. Many futurists believe each of our major body parts will likely one day be replaced or significantly modified by extreme science and technology.

Recently, I wrote about the growing use of artificial hearts and optimistically predicted in 10 years robotic hearts may be equal or better than human hearts. On September 5th, a revolutionary new robotic heart—the Carmat heart—was fitted into a patient in France. According to the AFP, French Health Minister Mariso Touraine said, "This intervention confirms that heart transplant procedures are entering a new era."

While researching artificial hearts, however, I was surprised to learn that healing damaged organs with stem cells or growing new organs outright may end up beating the creation of artificial organs to the marketplace.

Right away, I realized one of the great races of the 21st century might already be underway.

The technological rivalries of the 20th century have long fascinated people. The evolution of the transportation industry: cars vs. trains vs. airplanes. Or what about Bill Gates' infamous tactics against Apple and IBM to establish Microsoft Windows as the dominant operating system in the world?

The healthcare industry has rarely had such a publicized competition, as most innovation was isolated to its own specific causes and needs. But the modern world of invention is now more connected than ever.

Hardly any major new science or tech emerges without WiFi capabilities, or microchip processors, or something synthetic. For example, robotic limbs are connecting to the nervous system. Chip implants can release hormones in the body. And exoskeletons are helping the disabled walk.

The synergies of this interconnected world are stronger than ever. Coders, engineers, and biologists often work hand in hand, all in the same lab.

"There's no longer this big divide between medicine and technology," says Gray Scott, a futurist and co-executive producer of upcoming documentary The Future of Work and Death. "We are increasingly becoming cyborg-like beings. We are becoming literally what we create. Biology, physics, and technology are evolving towards one-and-the-same-thing."

So which approaches are worth watching in the race to repair and replace our organs? While there is no single dominant method yet, I'm keeping an eye on three technologies: robotics, stem cells, and 3D-printed organs.

Diehard robotics fans believe the future lies entirely in becoming machines. Indeed, I agree that the long-term future will be dominated by robotics and then later by mind uploading, where no biological organs are used at all.

But not everyone will sign up for such radical technology in their lives.

Some people, for spiritual or philosophical reasons and preferences, may not want such extreme transformation, especially in the near term, while it takes society time to adjust to the strange but inevitable transhumanist age.

In the next five years, those people may be able to have it all. New stem cell technology will likely outpace the creation of robotic body parts for human use.

In the UK, a major study of 3,000 patients is underway to determine if injecting stem cells into heart attack-damaged hearts will speed recovery and lengthen lives. Early results are positive, giving experts hope that injected stem cells (taken from patients' hips in the UK study) might offer a simple solution to healing damaged organs in the body.

In another independent study, damaged hearts in monkeys were given stem cell injections that also showed significant improved organ usage after the treatment. Also, doctors in Japan are beginning to test using induced pluripotent stem cells to fight macular degeneration. Induced pluripotent stem cells are produced from adult cells, unlike embryonic stem cells.

More fascinating is the possibility of growing or even 3D printing organs (or bioprinting as it's called) in the lab. For years now, scientists have engineered very minor parts of the body in laboratories, such as organ tissue. But now they have devised methods and devices that can create more complex cellular structures. For example, CNN reports that "last year a 2-year-old girl in Illinois, born without a trachea, received a windpipe built with her own stem cells." The trachea was first constructed with plastic fibers and then stem cells were later meshed with it to create the respiratory organ.

One company, Organovo, a California start-up, has also had success in 3D printing tiny strips of human liver. These may be used to test new drugs. But one day they hope to produce a fully functioning liver for commercial purposes. And a team from the University of Louisville in Kentucky has also had some success printing human heart valves and small veins earlier this year, according to AP.

If this all sounds a bit far-fetched, it's because it is so outside our realm of experience. Many people's immediate reaction to such radical technology is fear. However, experts think such tech and its acceptance in the future will be determined more by need than anything else.

People tend to cling onto their preferences until their health significantly deteriorates. In a life and death medical situation, most people opt to go with what will save their life and enable them to be healthy again.

Companies creating radical new medical tech, such as entire synthetic organs, are profoundly aware of this. They hope to sidestep some controversy by focusing on the human aspects of their innovations—such as how artificial organs might grant more time with grandchildren and loved ones—and not whether the organ was grown, printed, or created in a factory.

At the end of the day, for most people, being healthy, productive, and having the ability to spend time with loved ones is what remains most important. In this way, no matter what technology or field of science wins the competition to build the best organs—flesh, machine, or a mixture of both—we are all winners.

********

## 18) The Abortion Debate Is Stuck. Are Artificial Wombs the Answer?

Could an emerging technology reshape the battle lines in the abortion debate? Since Roe v. Wade was decided in 1973, that fight has been defined by the interlocking, absolute values of choice and life: For some, a woman's right to choose trumps any claim to a right to life by the fetus; for others, it's the reverse. But what if we could separate those two — what if a woman's choice to terminate a pregnancy no longer meant terminating the fetus itself?

That is the promise of artificial wombs, a technology that has already shown some success in tests on sheep fetuses. Early in a ewe's pregnancy, the lamb fetus is removed from her body and placed in a synthetic uterine environment in which it receives nutrients and fluids, and continues to develop to term, a process researchers call ectogenesis.

Artificial human wombs are still far in the future, and there are of course other ethical issues to consider. But for now, the technology is developed enough to raise new questions for the abortion debate.

In a 2017 issue of the journal *Bioethics*, two philosophers, Jeremy V. Davis, a visiting professor at the United States Military Academy at West Point, and Eric Mathison, a postdoctoral associate at Baylor College of Medicine, argue that while a woman has a right to remove a fetus from her body, she does not have the right to kill it. The problem is that, for now, the latter is inherent in the former.

Their argument builds upon that of the pro-choice philosopher Judith Jarvis Thompson, who famously argued in her 1971 paper "A Defense of Abortion" that women have a right to not carry a fetus for nine months — but that women do not have a right to be guaranteed the death of the fetus.

Such arguments point toward a disjunction in the abortion debate. Ectogenesis is the answer.

Synthetic wombs have an appeal far beyond the abortion debate, of course. They could revolutionize premature birth, which the World Health Organization calls the number one cause of death among children under 5.

The most advanced research in ectogenesis is underway at Children's Hospital of Philadelphia, where sheep fetuses have been removed from their mothers' bodies after 105 to 120 days — the equivalent, in a human, of 22 to 24 weeks — and placed in "biobags," clear plastic containers filled with amniotic fluid. So far the lambs have developed with few complications.

Biobag technology could be available for humans in as little as one to three years, according to Dr. Alan Flake, a fetal surgeon in charge of the Children's Hospital of Philadelphia artificial womb experiments. Another team performing ectogenesis research at the University of Michigan also believes they could have devices ready for humans in a similar time frame.

Some major supporters of artificial wombs are transhumanists, who believe in using technology to improve human health, intelligence and quality of life. Women's rights activists likewise support the research, aiming to free the female body.

But the promise of artificial wombs should appeal most to conservatives looking to reduce the 600,000 abortions performed annually in the United States, but pessimistic about the chance of overturning Roe any time soon. Every fetus that was going to be aborted but instead makes it into an artificial womb could be considered a life saved.

Dr. Daniel Deen, an assistant professor of philosophy at Concordia University in Irvine, Calif., recently said in an interview with the website Leapsmag: "If the technology gets developed, I could not see any Christians, liberal or conservative, arguing that people seeking abortion ought not opt for a 'transfer' versus an abortive procedure."

Obviously, the idea that science could short-circuit a moral debate is discomforting for some. As artificial wombs improve, biobags are likely to become a hot-button topic for conservatives, who will have to decide how far they want to use technology to accomplish their ethical goals.

There are practical challenges, too: Artificial womb transplants and births are sure to be dramatically more expensive than the typical 15-minute abortion procedure, which costs around $500. And if even a quarter of those fetuses that would have been aborted are brought to term artificially, 150,000 babies a year would be born, almost all of them likely to be put up for adoption — more than the total number of annual adoptions in the United States. Who will pay for those procedures, and who will care for those children once they are born?

It is unlikely that the abortion debate will be resolved soon — certainly not as a legal matter. But as a practical and philosophical one, artificial wombs offer a way for both sides in the debate to move forward. The only question is whether we are willing to accept the increasingly central — and beneficial — role that technology can play in resolving what were once considered immutable human problems.

*******

## 19) The Technology Transhumanists Want in Their Kids

Technology seems to be disrupting nearly every aspect of our lives. However, with the exception of toddlers thumbing their way through smartphone apps and watching Sesame Street on YouTube, raising children isn't significantly different than it was 20 years ago.

I would bet my right arm, though, that such gradual change won't be the case in another 20 years. The transhumanist age of child rearing is dawning. I discovered this last year when a pediatrician checked my 24-hour-old infant's hearing with a soundless brainwave headset. Just three years before, in the same hospital with my first child, infant hearing tests were being done using the decade's old beeping device that you physically stick in the ear and wait for the munchkin to react.

In general, when adults see new technology available to themselves, they are often more curious than skeptical. But when they think of new tech for their children, parents can become downright defensive. Curiosity no longer prevails, and protection mechanisms kick in strongly. Despite this, society is on a path to embrace an ever increasing amount of bizarre tech to be used in the raising of its children—including some things literally inside children.

In fact, the 20-year outlook is so radical, that it seems science fiction-like. For example, NBC forecast that many Americans will get "chipped" by the year 2017. Right now, the majority of implanted chips people have are for restoring hearing, and in some cases treating mental disease, such as epilepsy. But biohackers are increasingly implanting RFID (Radio Frequency Identification) chips into themselves. Even X Prize Chairman and Singularity University co-founder Peter Diamandis did this on stage recently during a speech.

I'm excited by the technology and am looking into getting this type of implant for my four-year-old daughter. As a US presidential candidate who is sometimes threatened on Twitter and other social media, I want more control on the whereabouts of my child. The new RFID chips are tiny and harmless, but they can be tracked on a smartphone.

Of course, they will be also be useful for when my daughter gets older and tries to play hooky from school. This might sound excessively controlling, but expect millions of parents to think this same way—especially those who sign on to some of the strict Tiger Mom philosophy for their children. In fact, in the future, I think the biggest celebration of one's teens will not be birthdays or graduations, but when parents give their kids the right to turn off the tracking chip they have inside themselves—or have it removed entirely.

Of course, by then, no one will remove implants. This is because the new generation of implants (which are being developed and will be here in the next few years) will be for much more than just tracking. They will be medical wonders that people will use to get daily updates via their smartphones about their bodies, including heart rate, temperature, and hydration.

Already today, though, there is some child rearing transhumanist technology that is being used. For parents, the first year of life for their child is often the toughest, filled with hourly challenges to make sure the baby doesn't kill itself accidentally. Sudden Infant Death Syndrome is the greatest fear, a general term for when an infant dies in its first year of life.

About 2000 kids die from SIDS in the US every year. But new wearable technology is already helping to fight that. A sock called Owlet worn by an infant collects heart rate, oxygen, and sleep data—and then sends it to a smartphone or the cloud for other devices to access. TempTraq, a flexible patch that is stuck on one's child monitors the temperature of an infant—again sending updates to your smartphone. The patch debuted this year in the Consumer Electronics Show in Las Vegas.

Also at CES 2015, Intel debuted a car seat clip that lets parents know if they've accidentally left their child in the car, something that happens far more than you might think in America.

And of course, so-called smart diapers, some made by popular Huggies, are already used. In them, a sensor lets you know when your child is wet and needs changing. One company Pixie Scientific, that successfully closed an Indiegogo campaign, is working on a diaper and smart phone app let you know about whether one's child is having a urinary tract infection, prolonged dehydration, and even kidney problems.

Some technologies completely remove danger from children entirely. What parent hasn't worried about their kid getting in a car with a drunk at the wheel? MADD, Mothers Against Drunk Driving, is one of America's most popular nonprofits, and has been a major influence in keeping people away from the wheel when inebriated.

But MADD may not be around much longer. Driverless cars will save the lives of tens of thousands of kids and adults every year the more prevalent they become on the road. In fact, it's more than likely my children will never learn to drive. They won't need to. Cars will be automated completely.

But just how far will technology go to changing child rearing? Currently, the conflict between home, public, and private schooling is a half-century old debate. Which is better and which will give your child the largest advantage later in its life? The debate is becoming even more complicated with online learning, where people can, for example, take MIT courses from their computers and get the world's best educations. Furthermore, virtual reality will make it even more complex as one will be able to participate in each of the three education environments.

Yet, it's another debate that could be obsolete in 20 years. Brainwave headsets or more advanced cranial implant technology could literally be the death of education. Using mind uploading tech, people in the future may be able to download educations and skills directly into their minds and memories.

Sound impossible? It's not. Already, last year telepathy was accomplished between two people across an ocean. Billions of dollars are being poured in the EGG, or mind wave reading and stimulating technology. Eventually, we'll find just the right tech and algorithms that enable us to download education directly into the learning parts of our brains. Then playing that Mozart's 5th Symphony won't take 10 years to learn, but 10 seconds to sync and start playing. Imagine how much time we we might save? Sure, our kids won't have much discipline or a love of learning new things, but they'll sure know French grammar (and Chinese, Russian, and Arabic perfectly too).

Of course, no conversation on transhumanism and children would be complete without talking of designer babies. Last month, Chinese scientists broke ground with CRISPR technology of editing and modifying the DNA of embryos. The fact is their experiment literally has helped usher in the age of designer babies, something society has long known it was going to enter. In just years, we might posses the ability to edit in higher intelligences to our offspring—and edit out hereditary diseases. Of course, we'll also be able to choose any eye, hair, and skin color, as well as other traits we might want in our children.

The controversy with this technology is two-fold. Will conservative or religious people let us remake the human being into a more functional version of itself? And will all people be able to afford it? Editing a genome isn't going to be cheap, at first. Neither will driverless cars. Furthermore, I surmise the Ivy League undergrad education download is also going to be costly (although, it'll probably still be much cheaper than a physical education). So, is all this transhumanist child rearing tech fair to those who can't afford it?

The short answer is: Of course, not. But neither are the costs of AIDS treatments in the world today. Hundreds of thousands still die because they can't afford the proper technology and medicine. And it's a fact that wealthier people live far longer, fuller lives than poor people—about 25 percent more on average.

So what can we do to even the playing field? To begin with, let's not stop the technology. Instead, let's work on stopping the inequality and create programs that entitle all children to better health and child rearing innovation. As a society, let's come up with ways that make it so all peoples can benefit from the transhumanist tech that is changing our world and changing the way our children will be raised.

********

## 20) A Brain Implant that Registers Trauma Could Help Prevent Rape, Tragedy, and Crime—So Why Don't We Have it Yet?

There's been a lot of talk across America about college rape culture in the last few weeks. Much major media has been highlighting the persistent and unfortunate problem. Perhaps the most well-known article came from *Rolling Stone*, which ran a highly controversial story highlighting seven University of Virginia fraternity students allegedly raping one freshman girl for hours during a party. In the wake of so much arresting coverage, numerous universities and legislative bodies are considering new methods to deal with the problem.

So far, those new methods seem to consist mostly of advocating for clearer language to stop the violence from happening in the first place and greater transparency in the rape victim's reporting process. I'm not optimistic the changes will do much to stop rape and other forms of criminal violence in any significant way. There are too many aggressive, idiotic men out there—and yes, men are almost always responsible for the violence.

The facts of domestic abuse in America are sobering. Nonprofit Arkansas Coalition Against Domestic Violence reports that every 15 seconds a woman is beaten and that 35 percent of all emergency room visits are a result of domestic violence. Nonprofit A.A.R.D.V.A.R.C., An Abuse, Rape, and Domestic Violence Aid and Resource Collection, reports that that the US Surgeon General states that domestic violence is the leading cause of injury to women between the ages of 15 and 44 in the United States. According to *Feminist.com*, over 22 million women in the United States have been raped in their lifetime, based on the National Intimate Partner and Sexual Violence Survey 2010. SafeHorizon, the largest victims' service agency in America, says there are 2.9 million reports of child abuse every year nationally, and it costs $124 billion dollars annually

64

in medical, court, law enforcement, child welfare, and juvenile protection services to deal with the problem.

As a transhumanist, I strive to consider all societal problems from technological and scientific points of views. It turns out there might be a simple solution that could reduce rape and some violent crime all across the country. I call it the trauma alert implant.

Cranial implants and brain wave technology—despite a Mark of the Beast reputation by Christian conspiracy theorists—have come a long way in the last few years. Already, hundreds of thousands of people in the world have microchip implants in their heads, consisting of everything from chips to help Parkinson's sufferers to cochlear implants for the deaf to devices to assist Alzheimer's patients with memory loss. For each, this technology allows a better life. DARPA recently announced a $70 million dollar five-year plan to develop implants that can monitor soldier's health. It's part of President Obama's new multi-billion dollar BRAIN Initiative.

Implants using Electroencephalography (EEG) technology can read and decipher brain waves. Trauma, however experienced, is a measurable biological phenomenon that can be monitored and captured by an implant device. Scientists must do nothing more than create a trackable chip that sends an emergency signal to nearby authorities when it registers extreme trauma. Help can then arrive quickly to the victim.

Much of the technology for such a device basically already exists. And such a device could be useful for far more than rape or criminal violence, too. Drowning, being burned in a fire, automobile accidents, building collapses, snake bites, kidnappings, bullet wounds, senior citizens who've fallen down stairs and can't get up— the list of terrible things that happen to humans goes on and on. The result of every one of them is almost always the same: brain waves that manifest extreme trauma—the human's most basic response and alert system. Regardless what misfortune happens to a human being, most experts agree that getting victims rapid emergency assistance is the single best way to help them.

Consider the 2-year-old boy was snatched away from its parents by an alligator at Walt Disney World in 2016. I have a similar-aged toddler myself, and I followed this heartbreaking story closely.

Unfortunately, it ended as horribly as it began, with the recovery of a dead child.

While scene reports claim the father got into the water to save his son, perhaps if that 2-year-old at Disney World had been GPS chipped, the parents could have tracked him on their smartphones. And security might have been able to quickly identify his location in the water, perhaps even fast enough to have rescued him. *The New York Times* reports that the body was tragically found underwater only 10-15 feet from where it was last seen.

That's of little consolation now, of course, and I don't mean to be insensitive to the family's loss, but I do think this tragedy illustrates how implants could help improve public safety. They could help track our children, and adults for that matter, in the case of kidnapping and Amber alerts, or even just when they get lost on a hike in the woods.

As the father of a 5-year-old who will be attending school next year, I'm a big believer in the future that all children will get chipped somewhere on their bodies, perhaps like all children get vaccines in the U.S. It's crazy to me that we don't develop and use this technology, especially with our children. I'm looking into getting my children chipped after this alligator incident and because, as a controversial presidential candidate, I have security issues myself to worry about.

Of course, it's not only implants. It's chip tattoos, GPS jewelry, wearable tech T-shirts, or even shoes with tracking tech built into them. Using tech to keep humanity safe is a burgeoning field. Interestingly, an industry already exists around children using tech to keep their whereabouts safe, but they're mostly children with disabilities—some who have a propensity to wander off.

Perhaps the most advanced case of chipping people already in existence has to do with the military. Reports describe special forces experimenting with them so they can be tracked. In 2014, for example, the U.S. Department of Defense announced a $26 million grant for a brain implant that would record, analyze, and potentially alter live electrical signals to soldiers. The military is getting so interested in implants that I was recently asked to consult for the U.S. Navy on research of chipping their service people.

Back to the trauma alert implant—which to me is the holy grail of safety. Another great thing about it is that not everyone would have to get it to help stop violent crime and domestic abuse. In fact, probably most people wouldn't (though, I surmise in the future many people will get one for a multitude of reasons). The existence of the chip itself—similar to a possible hidden camera in a room—would be enough to scare off many criminals, who would always be second guessing if their victim had one. This would especially be the case when it comes to crimes that are hard to prove or go habitually unreported, such as date rape.

All things considered, the trauma alert implant sounds like a sensible and impressive thing. So why don't we have them yet?

To begin with, Americans are wary of brain implants. They don't mind holding a cell phone to their ears for a half hour, but ask them to get a piece of sophisticated tech inside their heads and many freak out. They squawk how weird it is and that they don't want to be a cyborg (all the while spending untold hours surfing the internet, flying on jet airplanes 30,000 feet in the air, and taking multiple vaccines and pills). The transhumanist age is already here, whether it's weird or not. For most people, it's just a matter of culturally accepting it.

Another complaint that people have with implants is the privacy issue. Nobody wants to be trackable. Sure, that's understandable. But bear in mind, that every time you get on the internet, stop at a gas station, or use your credit card, you're already being tracked. We may all distrust surveillance, but that's not going to stop the gargantuan amount of cameras recording in America right now, many of them in public places. While solid information is hard to find on how many cameras are operating in America, Wikipedia reports that Chicago has at least 10,000, and the United Kingdom may have as many 4.2 million cameras, or 1 for every 14 people. The good news is, just like our cell phones we carry around, we'll likely have the option to turn off our implants anytime we want, thereby giving us control of who can watch us. Additionally, surely implants could be programmed so that they could "only" be tracked once they were triggered for extreme trauma.

But probably the most significant reason Americans dislike implants is because of religion. At least 80 percent of the country's population holds some form of faith—mostly of Christian denomination. And a significant number of Christian people consider brain implant technology to be the definitive Mark of the Beast—and a sign of End Times. I'm an agnostic tending towards atheism, so I don't understand those fears. But I do know that Revelations and the Second Coming of Jesus supposedly can't be stopped by people, according to the Bible, so perhaps Americans should work through their cyborg-phobias and embrace useful transhumanist technology. After all, if a gang of rape perpetrators suspected their victim had a trauma alert chip that would notify authorities, do you think they'd still commit the crime? Surely most wouldn't, especially not university students.

I'm grateful my young daughters will probably never have to worry about drunk drivers on their prom night. In a decade's time, most cars on the road in America will be driverless or come with alcohol detection systems that don't allow inebriated drivers. Such technological innovation is just a drop in the ocean of the benefits that progress brings to our world. The truth is that technology can help fix almost all the world's problems. It can also help with the tragedy of rape and criminal violence, which dramatically harms nearly 10,000 women and children a day in the US. If even a small portion of the population would have trauma alert implants, rape and criminal violence might be substantially reduced.

*******

## 21) The Era of Artificial Hearts Has Begun

Artificial Knees. Total hip replacements. Cataract surgery. Hearing aids. Dentures. We are a society bent on improving human health through substitution and augmentation of our body parts. But one of the most important goals of transhumanist medicine—possessing a perfectly healthy heart—has so far remained elusive. This week, we came a step closer when for the second time ever, a French company implanted a permanent artificial heart in a patient.

Heart disease is the number one killer in America: It claims nearly 800,000 lives every year, making it the cause of roughly one in three deaths. But radical new medical technology may soon change that: Expect the possibility of trading in your biological heart for a better, artificial one in about a decade's time.

There have been over 1,000 artificial heart transplant surgeries carried out in humans over the last 35 years. Probably the most famous is Dr. Barney Clark, a Seattle dentist who survived 112 days in 1982 with an implanted Jarvik-7, the first device designed to completely replace the heart. Over 11,000 more heart surgeries where valve pumps were installed have also been performed— former Vice President Dick Cheney received one in 2010.

Replacing the human heart with a robotic version is on the rise around the world. However, nearly all operations currently carried out are only a temporary bridge to buy precious time until a biological heart transplant can be made. Transplants of biological hearts, while often successful, are very difficult to come by, due to a shortage of suitable organs. Over 100,000 people around the world at any given time are waiting for a heart. Even Dick Cheney had to wait 20 months to find a heart appropriate for his body. There simply are not enough healthy hearts available for the thousands who need them.

This shortage has prompted numerous medical companies to jump into the artificial heart game, where the creation of a successful and permanent robotic heart could generate billions of dollars and help revolutionize medicine and health care.

Last December, one such company took a giant leap ahead. French-based Carmat performed the world's first total artificial heart implant surgery on a 76-year-old man in which no additional donor heart was sought.

Carmat—led by co-founder and heart transplant specialist Dr. Alain Carpentier—spent 25 years developing the heart. The device weighs three times that of an average human heart, is made of soft "biomaterials," and operates off a five-year lithium battery. The key difference between Carmat's heart and past efforts is that Carmat's is self-regulating, and actively seeks to mimic the real human heart, via an array of sophisticated sensors.

Unfortunately, the patient who received the first Carmat heart died prematurely only a few months after its installation. Early indications show there was a short circuit in the device, but Carmat is still investigating the details of the death.

On September 5th, however, another patient in France received the Carmat heart. "This intervention confirms that heart transplant procedures are entering a new era," French Health Minister Marisol Touraine said Monday in a statement, according to the AFP.

Naturally, some critics worry that, beyond the efficacy of the device itself, an artificial heart is too large a step towards becoming a cyborg—a term the public still isn't yet comfortable with. However, many futurists, like myself, believe such transhumanist tech won't scare most people in the long run.

When you talk about replacing an arm—another technology that is already here and may become elective in as soon as 10 years—people freak out. They are not ready to see themselves in the mirror or at the beach with a metal or rubbery prosthetic, even if functionally, it's actually better than their original arm. But the artificial heart will be hidden inside the body, and it will soon be better than the heart of any Olympian athlete.

I surmise that millions will electively line up for it when it becomes available, even if they have healthy biological hearts. The biggest dilemma with the heart will probably be affordability. Currently, the Carmat heart costs about $200,000.

More than just pumping blood, future artificial hearts will bring numerous other advantages with them. They will have computer chips and wi-fi capacity built into them. We'll control our hearts with our smart phones, tuning down its pumping capacity when we want to sleep, or tuning it up when we want to run marathons.

The benefits could be endless. Have you ever been super nervous—such as on a first date, or while giving a public speech—and could feel your biological heart incessantly pounding? In the future, you'll just adjust your artificial heart to the right level for whatever context or experience you are in. Want to meditate? Turn the artificial heart to Buddha mode. Want to emulate a porn star? Turn it up for wild sex.

Of course, health experts would probably find this problematic. For example, our heart responds to its surroundings for a reason, and overriding such stimuli may cause unintended consequences. The body and its other organs may not be able to keep up with a fired-up 65-year-old who has purposely sped up his heart in order to surf giant Hawaiian waves.

Future artificial hearts may also replace the need for some doctor visits and physicals. Every second of the day, the device will monitor your health and blood, and relay updates of oxygen levels, whether you've contracted HIV, or if your alcohol content is too much for driving. In fact, much of that cool wearable medical tech that is in vogue right now, like the Fitbit Flex, will likely become obsolete once the artificial heart arrives in its perfected form.

One major downside of artificial hearts is their exposure to being hacked. Imagine the chaos that the mafia, an authoritarian government, or malignant hackers could cause if one's artificial heart were targeted. Viruses could be sent into the heart's software, or the password to the app controlling your heart could be stolen and misused.

Dick Cheney was so worried about his implanted heart defibrillator being attacked by terrorists during his vice presidency that he asked to have its wi-fi capabilities removed, according to his book, *Heart: An American Medical Odyssey.*

Of course, the hacking of hearts already happens in a metaphorical way, whether by malice, love, or chance. Just think of your first passionate romance, or your dearest family members. My young daughters have long held the keys to my heart.

At a recent press conference, Carmat's Dr. Carpentier put it differently, repeating the words of the famous French poet Calude Bernard: "Whatever the poets may say, the heart is just a pump."

*******

## 22) The Coming Genetic Editing Age of Humans Won't Be Easy to Stomach

Some futurists believe humans will eventually become all ones and zeroes, a result of a total merger with machines and the microprocessor, before this century is out.

Standing in the way of this are older religious humans who overwhelming control governments and legal policy around the world, and they will insist we remain biological mammalian entities for as long as possible.

One could argue, however, that the coming Star Wars-like age of speciation—as widely seen in a rough bar on planet Tatooine—will challenge our mental outlook on the human form far more than machines.

Right now, the body transformations humans undergo seems harmless to most people. Even conservatives shrug at typical modifications: pierced noses, magnets in finger tips, and implants in our forehead to make it appear like some humans have devil-like horns.

In fact, a mostly accepting culture of synthetic parts and body modifications has already partially been built into modern medicine. Dentures don't scare us. Getting artificial hips when needed are a no-brainer. And even small implants in our hands don't worry us too much (I have one).

But these are nothing compared to what biohackers want to do in the near future. Some want to grow a third eye on the back of their head—a feat which isn't as complicated as it sounds and could happen in as little as five to 10 years. Some of this tech is already here. New gene editing technology, such as CRISPR techniques—where scientists cut and edit human DNA to affect human biology—has already produced dogs with larger muscles. Another CRISPR-like technology called TALEN has been used to eliminate cancer from a child.

In the future, probably not too many people will mind genetic editing that makes us taller, or changes the color of eyes. And even fewer people will disagree with using this type of science to eliminate hereditary disease, such as Alzheimer's or Diabetes. But what about growing an extra set of blue colored arms like the Hindu lord Krishna? Or what about growing a horse's lower body so humans can become centaurs? I've even heard male biohackers talk about trying to grow a second penis right above their primary one.

Immediately, these ideas make many people cringe. I call this unease *speciation syndrome*, where witnessing significant physical genetic transformation of human beings causes revulsion and shock. It can also happen to those who undergo the transformation themselves.

Despite initial unease to major bodily modification, I like the idea of having an extra eye on the back of my head—or another set of limbs. Or even a pair of wings. Some of it can be quite functional. However, even an extra eye on the back of one's head is likely to be shocking and possibly terrifying for most people. Can you imagine the first person that gets one? He or she is likely to become known as the weirdest person in the world.

But what exactly is it that freaks us out? Why is it an issue to have our physicality dramatically altered? To answer that question, let's first look at speciation syndrome's cousin concept. In technology circles, it's known as the Uncanny Valley, where a robot that becomes too humanlike makes us feel unease or even revulsion. In fact, the more humanlike the machine becomes, the worse we generally feel.

The Uncanny Valley concept gives us some insight into the complexities of the human mind and its resistance to change. I surmise with speciation syndrome humanity will also discover its sense of limits to what genetic editing means for future human form—and this discovery will probably ultimately cause revulsion at first. After all, some people have a visceral reaction to unusual appearing humans, something I've witnessed firsthand in Cambodia's Killing Fields, where physically deformed people and limbless war victims openly feature their physical differences in order to make more money begging from tourists.

Transhumanism tech like CRISPR, 3D printing, and coming biological regeneration of limbs will not only change lives for those that have deformities, but it will change how we look at things like a person with a three-foot tail and maybe even a second head.

At the core of all this is the ingrained belief that the human being is pre-formed organism, complete with one head, four limbs, and other standard anatomical parts. But in the transhumanist age, the human being should be looked at more like a machine—like a car, if you will: something that comes out a particular way with certain attributes, but then can be heavily modified. In fact, it can be rebuilt from scratch.

In the future, there may even be walk-in clinics where people can go to have various gene treatments done to affect their bodies. Already, we have IVF centers where people can use radical tech to privately get pregnant—and also control and monitor various stages of a child's birth. Eventually, if government allows it, gene editing centers will also offer a multitude of designer baby traits, some which also would come via CRISPR. We might even eventually use artificial wombs for the whole process.

Economically, a trillion dollar industry could be created by the burgeoning genetic editing industry—one that greatly benefits human health and science innovation. But of course, first we must get over our fears of modifying the human body and the effects of speciation syndrome.

The best way to get society over that original hump is to focus on and praise CRISPR's ability to wipe out disease. We might even eventually be able to eliminate aging via coming genetic treatments. However, before we start adding arms and extra eyes to our bodies—something I support and look forward to doing myself someday—I hope scientists will bring about socially acceptable ways to live longer and stop disease with these amazing new techniques. That way speciation syndrome may not be so uncanny afterall.

********

# CHAPTER IV: CRYONICS

## 23) Can Cryonics, Cryothanasia, and Transhumanism Be Part of the Euthanasia Debate?

An elderly man named Bill sits in a lonely Nevada nursing home, staring out the window. The sun is fading from the sky, and night will soon cover the surrounding windswept desert. Bill has late-onset Alzheimer's disease, and the plethora of medications he's on is losing the war to keep his mind intact. Soon, he will lose control of many of his cognitive functions, will forget many of his memories, and will no longer recognize friends and family. Approximately 40 million people around the world have some form of dementia, according to a World Health Organization report. About 70 percent of those suffer from Alzheimer's. With average lifespans increasing due to rapidly improving longevity science, what are people with these maladies to do? Do those with severe cases want to be kept alive for years or even decades in a debilitated mental state just because modern medicine can do it?

In parts of Europe and a few states in America where assisted suicide—sometimes referred to as euthanasia or physician aid in dying—is allowed, some mental illness sufferers decide to end their lives while they're still cognitively sound and can recognize their memories and personality. However, most people around the world with dementia are forced to watch their minds deteriorate. Families and caretakers of dementia patients are often dramatically affected too. Watching a loved one slowly lose their cognitive functions and memories is one of the most challenging and painful predicaments anyone can ever go through. Exorbitant finances further complicate the matter because it's expensive to provide proper care for the mentally ill.

In the 21st Century—the age of transhumanism and brilliant scientific achievement—the question should be asked: Are there other ways to approach this sensitive issue?

The transhumanist field of cryonics—using ultra-cold temperatures to preserve a dead body in hopes of future revival—has come a long way since the first person was frozen in 1967. Various organizations and companies around the world have since preserved a few hundred people. Over a thousand people are signed up to be frozen in the future, and many millions of people are aware of the procedure.

Some may say cryonics is crackpot science. However, those accusations are unfounded. Already, human beings can be revived and go on to live normal lives after being frozen in water for over an hour. Additionally, suspended animation is now occurring in a university hospital in Pittsburgh, where a saline-cooling solution has recently been approved by the FDA to preserve the clinically dead for hours before resuscitating them. In a decade's time, this procedure may be used to keep people suspended for a week or a month before waking them. Clearly, the medical field of preserving the dead for possible future life is quickly improving every year.

The trick with cryonics is preserving someone immediately after they've died. Otherwise, critical organs, especially the brain and its billions of neurons, have a far higher chance of being damaged in the freezing. However, it's almost impossible to cryonically freeze someone right after death. Circumstances usually get in the way of an ideal suspension. Bodies must first be brought to a cryonics facility. Most municipalities require technicians, doctors, and a funeral director to legally sign off on a body before it can be cryonically preserved. All this takes time, and minutes are precious once the last heartbeat and breath of air have been made by a cryonics candidate.

Recently, some transhumanists have advocated for cryothanasia, where a patient undergoes physician or self-administered euthanasia with the intent of being cryonically suspended during the death process or immediately afterward. This creates the optimum environment since all persons involved are on hand and ready to do their part so that an ideal freeze can occur.

Cryothanasia could be utilized for a number of people and situations: the atheist Alzheimer's sufferer who doesn't believe in an afterlife and wants science to give him another chance in the future; the suicidal schizophrenic who doesn't want to exist in the current world, but isn't ready to give up altogether on existence; the terminally ill transhumanist cancer patient who doesn't want to lose half their body weight and undergo painful chemotherapy before being cryonically frozen; or the extreme special needs or disabled person who wants to come back in an age where their disabilities can be fixed.

There might even be spiritual, religious, or philosophical reasons for pursuing an impermanent death, as in my novel *The Transhumanist Wager*, where protagonist Jethro Knights undergoes cryothanasia in search of a lost loved one.

There are many sound reasons why someone might choose cryothanasia. Whoever the person and whatever the reason, there is a belief that life can be better for them in some future time. Some experts believe we will begin reanimating cryonically frozen patients in 25 to 50 years. Technologies via bioengineering, nanomedicine, and mind uploading will likely lead the way. Hundreds of millions of dollars are being spent on developing these technologies that will also create breakthroughs for the field of cryonics and other areas of suspended animation.

Another advantage about cryonics and cryothanasia is their affordability. It costs about $1,000 to painlessly euthanize oneself and an average of $80,000 to cryonically freeze one's body. It costs many times more than that to keep someone alive who is suffering from a serious mental disorder and needs constant 24-hour a day care over many years.

Despite some of the positive possibilities, cryothanasia is virtually unknown to people and is often technically illegal in many places around the world. Of course, much discussion would have to take place in private, public, and political circles in order to determine if cryothanasia has a valid place in society. Nevertheless, cryothanasia represents an original way for dementia sufferers and others to consider now that they are living far longer than ever before.

********

## 24) I Visited a Facility Where Dead People are Frozen so They can be Revived Later

Over 100,000 people die each day globally. Why don't more of us consider cryonics — the practice of freezing the clinically dead in the hopes of bringing them back to life at a later date — as a way to avoid death?

As part of my 2016 US Presidential campaign representing the Transhumanist Party, I had a chance to stop in at Alcor Life Extension Foundation, the world's best-known cryonics facility, to find out.

You've probably seen cryonics before. Hollywood loves to use it in movies. Mike Myers (as Austin Powers), Woody Allen, and Mel Gibson are just some examples of people who have been "frozen" on the big screen.

Cryonics — and the field of life extension — has also been in the news a lot lately. A recent *New York Times* article featuring an Alcor patient generated discussion across the Internet.

Most cryonicists would not call frozen patients dead. They say patients are temporarily beyond the help of modern medicine, and that cryonics is the final attempt to provide emergency healthcare. Cryonics, they argue, is actually saving a patient by buying them time for science to catch up to the point where they can be revived.

"Death is not a moment, but a process, where an individual goes from a state of health, through many steps which end up becoming irreversible by modern means. It is not an absolute event. It is almost entirely dependent on the skills and means of the rescuer, and as we know, those skills and means improve over time," says Christine Gaspar, a RN and President of Cryonics Society of Canada and CEO Biostasis Canada.

I visited Alcor while traveling cross-country aboard my Immortality Bus (a campaign bus that resembles a coffin). Dr. Max More, a philosopher and the CEO of Alcor, gave me and my Transhumanist

Party volunteers a private tour at the nonprofit's Scottsdale, Arizona facility.

One thing that struck me about Alcor was its size. It's not a small shop housing a few dead people in big steel tubes. It's a giant medical facility, complete with offices, surgical bays, laboratories, conference rooms, and of course, a large, highly secured hall for the cryonic tanks, known as dewars.

More oversees many of the cryonics procedures, and has a medical and scientific advisory board to look after operations. His team includes medical doctors, paramedics, and surgeons. 138 patients have been placed in cryonic suspension at Alcor so far.

Among these patients are baseball legend Ted Williams, transhumanism advocate FM-2030, and James H. Bedford, PhD., the first patient to be cryopreserved back in 1967.

"FM-2030 is a good friend of mine," More tells me.

That comment made me wonder about the immense challenges and commitment of being responsible — literally —for the existence of one's good friends. It seems overwhelming, but More, a fit 51-year-old, is up to the challenge. He is a steward for the transhumanist community — overseeing the bodies of friends and their families as they grow too old for science to help them. He acts as their guardian, advocate, and spokesman.

The process of cryonics begins with signing up for the service and gaining membership at Alcor or one of the other few cryonics facilities in existence. Ideally, a patient dies near a cryonics facility, so that they can be immediately cooled and prepared.

This is the ideal condition for diffusing cryoprotectants in the brain: Cooling the patient down to liquid nitrogen temperatures in the most controlled method possible, so that brain neurons containing memories and (hopefully) identity can be protected and preserved.

Many cryonicists wear "dog tags" or other identifying jewelry that show they require cryopreservation immediately after pronouncement of death — and medical professionals are supposed to respond to that. Dr. More told me one person even had instructions tattooed on himself so that they could be easily seen. Patients that aren't transported to a cryonics facility within a few hours of death are thought to not be preserved in ideal conditions.

Patients are placed in a bath of ice for transport and infused with chemicals to help preserve their cells and tissue structures in a process called vitrification. This, hopefully, eliminates the formation of ice crystals that can puncture cell walls and destroy the cells themselves. Later, either the head or whole body (depending on the preference of the patient) is transferred into a giant dewar filled with liquid nitrogen. Preserving just your head at Alcor is about half the price of the body, coming in around $80,000 plus minimal annual fees.

The science at Alcor and in cryonics is constantly improving, according to More. He tells me the new techniques they've been using the in the last 10 years are better at staving off the ice crystals that scientists suggest might lead to damage of the brain.

When I first began my Immortality Bus tour, I considered transforming the bus into a cryonics dewar to raise attention to life extension issues. But so few people knew about cryonics that the bus's effect on the public would be muted. So I chose to make my bus look like a coffin, and most people get it right away.

Because I'm in excellent health, and cryonics is expensive, I haven't signed up yet. However, More told me that life insurance can help provide financial means to get the cryonics procedure done. Now I'm set on signing up for cryonics before my presidential campaign ends, in hopes of bringing attention to this small but potentially life-changing industry.

When I asked More why more people don't sign up for cryonics, he shrugged and said, "I don't really know. You would think everyone that likes living would be interested in this. But so far few people have signed up."

As an aspiring politician, I advocate for government policy that specifically protects citizens' lifespans. Today, cryonics is the only hands-on treatment I know of that has a shot a preserving the lives and minds of the people we love. If people could become more comfortable with the idea of cryonics as emergency medicine rather than simply "freezing the dead," then I think it might become a much larger industry.

\*\*\*\*\*\*\*

## 25) Cryonics, Special Needs People, and the Coming Transhumanist Future

Recently, I was at Peet's Coffee writing an article on my laptop. A tired father walked into the shop with his adult son, a portly-looking 20-year-old weighing over 200 pounds. The son had Down syndrome, and his mental state was so confused that the father had to walk closely behind him, holding both of his shoulders to guide him. The son moaned as he walked, jerking forward in sharp, uncoordinated movements. Saliva bubbled out of his mouth.

I'm the parent of two young children (a 3-year-old and an 11-week-old infant), and my sympathy immediately went out to this father and his grueling burden in life. For many parents—especially a nonreligious one like myself—having an extreme special needs child is a daunting worry. Every five minutes in America, a child is born mentally retarded. That's over 100,000 kids a year. Approximately three percent of the American population has some form of severe cognitive dysfunction.

I watched the father place his order with the Peet's barista, receive his coffee, and lead his son to the condiment bar right next to me. The father released his son for a moment while he put creamer into his coffee. Within two seconds, the son arbitrarily lunged for my tea, spilling it all over my computer. He then proceeded to the next table and did the same with their drinks, yelling and grunting riotously.

Many people, including myself, jumped up and helped the father regain control of his son. It took only one look at the father's moist eyes to see how difficult this man's life was—filled with endless public apologies for his son's unpredictable behavior.

The question society must ask itself in the 21st Century—the age of transhumanist science and technology: genetic engineering, cyborgism, artificial intelligence, robotics, and radical life extension research—is what is the best way to handle such extreme special needs people? There are over seven million people in America with severe learning and cognitive disabilities—the most common are Down syndrome, Velocariofacial syndrome, and Fetal Alcohol syndrome—of which about 250,000 are institutionalized. Many of them are in poor health and will die prematurely. For both parents and society, the obligation is both massive and challenging. It costs many billions of dollars to keep extreme special needs people alive and not hurting themselves or others.

The field of cryonics—where human bodies are frozen using ultra-cold temperatures—has come a long way since the first person was preserved in 1967. Organizations like Alcor, Cryonics Institute, and Suspended Animation Inc. have together frozen ("suspended" in cryo-talk) a few hundred people. Eventually, science will figure out a way to bring them back to life—to revive them. Already, some people have survived death in freezing water for over an hour and have been brought back to life. Additionally, new techniques using a saline-cooling procedure can help restart the lives of people who have been recently declared clinically dead. Each year science advances, and the chances for reanimation of cryonic patients improve.

Given the feverish pace of scientific growth and innovation in the modern world, would it not be better to cryogenically freeze severely mentally retarded people with the hope of bringing them back to an age where science can cure them of their imperfection? More so, is it moral in the 21st Century to allow them to exist and die as they are, when likely in a matter of decades science will have what it needs to genetically alter them and make them cognitively normal? Don't we owe them the chance to be like us?

As a transhumanist philosopher, I advocate going further than just preserving special needs people after they die. I believe parents should have the legal right to painlessly put their extreme special needs children into a cryogenic state while they are biologically healthy and have years left on their lives. Some extreme special needs people are clearly unhappy, living in a nightmarish rollercoaster mental state—one that is also excruciatingly painful and crushing for their families. The all-important question to ask is: If it was you in their position—either as the parent or the special needs person—what would you want? The answers, at least for the nonreligious, are quite obvious.

So why then is this act illegal? Why is society afraid of evolving its perspective on this? Is it religion? Cultural stigma? Or are we simply lazy and prefer turning a blind eye to the controversial matter?

Many may say cryogenically preserving someone while they're still biologically healthy is murder (since it would technically involve stopping their life to successfully complete the process), but what do you call a person persistent on enslaving someone in decades-long confusion, insanity, and possible suffering when a more reasonable option exists? More importantly, like many other transhumanists and life extensionists, I no longer believe in death when it involves cryonics. Cryonics is more similar to sleeping or hibernation: a machine with the power button temporarily switched to "off." This is the 21st Century; dying is going the way of the dinosaurs. If you don't believe it, you're not reading scientific and medical journals. Additionally, you're not talking to trauma surgeons who can preserve technically deceased gun-shot victims for hours at a time before bringing them back to life.

Human civilization is at the cusp of achieving indefinite life extension for our species. Many leading bio-gerontologists say it's only a matter of decades before we can stop or reverse aging in people. Experiments are already succeeding with this in mice. Furthermore, hundreds of millions of dollars are being poured into genetic engineering research. Additionally, replacing old body parts with new artificial body parts will become commonplace in five to ten years. Perhaps most immediately promising, the use of stem cells to rehabilitate disease and malfunction in the brain is already being used with some success in research laboratories and hospitals. Clearly, if we can just get extreme special needs people to live long enough—or we can cryopreserve them if parents prefer—we will have a chance in the future to make them cognitively normal.

Currently, the situation today with extreme special needs people is anything but *normal*. While some are institutionalized and cared for by the state, many others are not. Some families make the choice to care for their special needs members. This is, of course, incredibly difficult to do and often leaves everyone miserable. Besides enormous time and financial loss, there are immeasurable emotional tolls. Marriages sometimes break up over attempting to provide the care. Healthy and intelligent siblings are regularly given the cold shoulder due to the constant demands of special needs siblings. Attempting even the most basic public outings with a special needs person (such as getting coffee at Peet's) can become a dangerous, complicated ordeal. The list of negative repercussions for anyone trying to provide care for a mentally retarded person goes on and on.

One of the main reasons I advocate cryonics as a possible consideration for severe special needs people—whether they're in the middle of their life or the end of it—is for the parent's sakes. Wouldn't parents rather live happy, productive, and liberated lives rather than spending their time changing diapers and spoon feeding an unruly adult with 20 more years to live? And by the way, that extra 20 years is actually going to be an extra 50 years in a decade's time given how fast life extension science is advancing.

Another point to consider are the financial aspects of cryonics for severe cases of special needs people. Cryogenically preserving someone costs approximately $150,000, and then approximately $1000 annually in maintenance and storage fees. Caring for a special needs person runs at least $1,000,000 over their lifetime according to a US Government CDC report in 2003, and that figure is likely much higher in severe cases, especially considering the increasingly length of lifespans due to modern medicine. (Of course, one also needs to add in 11 years of inflation from when that 2003 CDC report was published too.) In short, it's probably dozens of times cheaper to go the cryonics route.

To honor society's commitment to special needs people that are cryogenically preserved, we could put all the resources we were going to spend on their lifetime care into cryonic rejuvenation science and technology, as well as  genetic engineering which in the future will likely be able to reverse mental retardation. That amount of redirected money would equal many billions of dollars. In return, such newly funded science would also help future-born people with special needs, as well as the human population as a whole. Crossover science would certainly occur and spur new technologies, medicines, jobs, and ideas.

The future is coming far more quickly than most people realize. Bionic arms now connect to human nervous systems. Computer chips are already being put in people's heads for medical reasons. Video games can be played using just brainwaves. Genetically engineering humans will become a reality in just a few years. Artificial general intelligence, the holy grail of technology that may solve many of humanity's problems, will arrive within a decade or two. With such incredible technology and science at our species' disposal, an entire new set of rights and wrongs, as well as moral ambiguities, will challenge us all. As I mentioned before, the most pertinent question one can ask when facing such radical transhumanist technology is this: If it were me or if it was my child that could benefit from such advances and ideas, would I endorse it? In the case of the young man in Peet's afflicted with severe Down syndrome, who likely has decades left of his life, the controversial proposition of cryonic suspension should be considered. Anything else for him or the hundreds of thousands of others like him represents missed opportunity, possible injustice, and maybe even profound inhumanity. As a society living in the 21st Century, we can do better for the special needs people that deserve a chance at becoming normal.

******

## 26) 'Dead is Gone Forever:' The Need for Cryonics Policy

The news that a 14-year-old U.K. girl has undergone cryonic suspension after dying from cancer has created headlines around the world, putting a spotlight on the need for cryonics regulation.

Cryonics—also sometimes called cryogenics—uses ultra-cold temperatures to freeze a human body in nitrogen in hopes that future technology and science might revive the person to health again. With so much radical medical innovation occurring at breakneck pace—like genetic editing, creation of synthetic cells and nanotechnology— cryonicists are hopeful that within 35 to 50 years, the first frozen patients may be revived.

I spent the last few days in Seoul, Korea, speaking at the Global Leaders Forum 2016 on life extension science, of which cryonics is a core part. One of the other attendees was Max More, president and

CEO of Arizona-based Alcor, one of the leading cryonics facilities in the world. Last year, More gave me a tour of Alcor, and I saw for myself the 8-foot-tall steel dewars where patients are preserved.

I say patients, because no cryonicist believes they are dead. They are suspended—think Hans Solo in Star War's Return of the Jedi—and waiting to be "reanimated." The lingo may sound odd, but there are thousands of cryonicists around the world and decades of scientific research backing up the science.

The story of the 14-year-old girl, who died in October in the United Kingdom and was cryonically preserved in America, was historic because it was the first court case of its kind ever to be brought up in England or Wales—or perhaps anywhere in the world. The mother of the 14-year-old wanted her daughter preserved, as did the 14-year-old herself. But the father didn't want it. A court case ensued where Judge Peter Jackson visited the dying girl and listened to her wish to attempt to live longer via cryonics.

In a letter to the judge, the 14 year-old wrote:

"I have been asked to explain why I want this unusual thing done. I am only 14 years old, and I don't want to die but I know I am going to die. I think being cryo-preserved gives me a chance to be cured and woken up—even in hundreds of years' time.

"I don't want to be buried underground. I want to live and live longer and I think that in the future they may find a cure for my cancer and wake me up. I want to have this chance. This is my wish."

Right before the teenager's death, Jackson ruled it was okay for her to use cryonics to be preserved. According to the BBC, " The girl's solicitor, Zoe Fleetwood, told BBC Radio 4's Today programme it had been a "great privilege" to be involved with the case of an "extraordinary individual."

The case sets new precedent for cryonics policy and regulation, something which I advocate for but is desperately lacking around the world. Some countries such as France and parts of Canada outright criminalize cryonics, arguing that the state has rights to a deceased body, but not the actual person.

Bill Andrews, a supporter of cryonics, biologist, and president of life extension company Sierra Sciences, says, "In the 21st century, government must step up and create policies that help those that want to live far longer—and that means creating a legal framework for cryonics to thrive in."

I think not only should cryonics be allowed all over the world, but governments must take steps to ensure we respect the rights of patients' bodies—not matter what they want to do with it. The age of radical life extension is upon us, no matter what form it takes. Already, since the start of the 20th century, lifespans have doubled in the developed world. Increasingly, some scientists think we might end aging within 25 years via science.

Approximately 150,000 people on planet Earth die every day. To those who don't want to die, but disease or aging has caught up with them, cryonics offers the best solution in the near term to stave off death and be given a second chance.

*******

## 27) Civil Rights Clash: Transhumanists Prepare to Challenge an Anti-Cryonics Law in Canada

British Columbia, the westernmost province of Canada that is home to over 4 million people, has a law on the books that prohibits the marketing or sale of services for cryonics. You can still deep-freeze people after they die in the province, but as you might guess, without advertising, it makes it nearly impossible to find willing funeral directors, doctors and technicians to pull it off — all which are required by law when dealing with a dead person. The surging global popularity of the transhumanism movement — which cryonics is an integral part of — is bringing pro-science activists out of the Canadian woodwork. The stage is being set for a civil rights clash in British Columbia. Cryonicists say they will challenge the law in court, citing it as a human rights violation that threatens their ultimate transhuman goal of trying to live indefinitely.

How this strange anti-science law of cryonics — part of the broader, well-researched field of cryogenics — ever got passed is murky. In 1990, a small group of bureaucrats quietly passed the law, the first of its kind in the world. Transhumanists complain there wasn't a single cryonicist on the board of regulators who championed the new law. Politicians and regulators say the law was passed to protect people from falling for vague, unprovable ideas that promise a reanimated life after death, whereas no such science exists yet to do that. But Ben Best, former President of the Cryonics Society of Canada (CSC), once made dozens of phone calls in the 1990s and publicized his conversations in a document that suggests another important theme was religious opposition to cryonics.

Indeed, the very nature of cryonics — where human bodies are preserved using ultra-cold temperatures so they can be brought back to life at a later point — flies in the face of every major Abrahamic religion, including Christianity. Many of the funeral directors in the British Columbia Funeral Association are religiously oriented or provide religious services. Clearly, this suggests a conflict of interests, especially since many cryonicists hold atheist or agnostic perspectives.

Cryobiology ambitions in humans began in North America in the Unites States in the early 1960s with the publication of academic Robert Ettinger's book The Prospect of Immortality. The field of cryonics has come a long way since the first person was preserved in 1967. Various organizations and companies around the world, like Alcor, Cryonics Institute, and Suspended Animation, Inc., have since frozen ("suspended" in cryo-talk) a few hundred people. The costs vary widely, but it's between $12,500 and $200,000 to freeze either a head or a whole body.

Because science advances each year, and the chances for reanimation of cryonic patients is always improving, it seems odd that British Columbia would uphold an obvious anti-science law. However, years of lobbying and letter writing from cryonicists have done little to change the minds of politicians. Ultimately, with more and more people supporting cryonics and transhumanism, a showdown is inevitable.

Members of the Cryonics Society of Canada and other supporters, in collaboration with a civil rights attorney, are preparing to challenge the law. It's the lawyer's opinion that the best way to challenge it is to start a cryonics company that markets its services in direct violation of the law, forcing the issue into court. Canadian cryonicists have received some funding for this already from Florida-based Life Extension Foundation, but more financial support will be needed in the future. The name of the proposed company is Biostasis Canada. Its purpose will be to provide services to help cryonics patients get the critical field stabilization procedures that are vital to good quality cryopreservation before they are transported to a cryonics facility in the US.

Christine Gaspar, a longtime cryonics supporter and an emergency nurse who spent four years in the Canadian Armed Forces, has been asked to be CEO. Over email, she told me:

We are creating Biostasis Canada. Its intent is to open shop in British Columbia and offer field perfusion service. It will likely spend the first year or two of its life in court challenging the law. Once that is accomplished, we'll aim to expand to provide service nationally. This will raise the bar for quality cryopreservation for Canadians.

If Biostasis Canada fails in its goals, more is at stake then just a small startup. Transhumanists are worried that if they can't defeat this law, it could set a precedent for other territories to create their own regulation. Already, the province of Alberta has considered a similar law, and, as of 2002, France has prohibited cryonics nationwide. The near future of transhumanism, with people born today expecting to live to age 150, artificial general intelligence on the horizon, and robots taking jobs away is a political minefield. Indubitably, the next few decades will be dominated by how science and technology are changing our species, and how far society will let that process occur.

Despite sounding far-fetched, many transhumanists do not intend to remain human at all. For them, the future of becoming cyborgs or totally merging with machines is what matters. They want to leave behind their flesh, which many transhumanists consider fragile and imperfect. In the meantime, preserving that flesh is the only way to make it to the future, which is why cryonics is such an important piece of the transhuman puzzle that must be defended at all costs.

For transhumanists all around the world, cryonics is a matter of the highest importance — of life and death, of existence and nonexistence. It's possible, given how fast science is improving, that reanimation of patients could begin occurring in as little as 20 to 30 years. Technologies via bioengineering, nanomedicine, and mind uploading will likely lead the way. Hundreds of millions of dollars are being spent on developing the technologies that will also create breakthroughs for cryonics. Word is getting out about that too. A recent study found that over half of Germans know what cryonics is, and one in five of them would consider using cryonics. That's no small number for a leading nation of science with 82 million people.

In a world where over 90 percent of the people hold religious views of the afterlife, cryonics could become a noteworthy global civil rights issue. Regardless what happens in the future, transhumanists consider anti-cryonics laws a serious violation of people's freedoms. Plenty of pro-science people around the world feel the same way too. They say they feel this way because they know the standard definition of human death is rapidly changing in the 21st Century. For example, suspended animation is now occurring in a university hospital in Pittsburgh, where a saline-cooling solution has recently been approved by the FDA to preserve the clinically dead for hours before resuscitating them. This is not science fiction. This is happening today. In a decade's time, that saline-cooling procedure may be used to keep the clinically dead deceased for a week or even for a month before they are brought back to life.

Eventually, modern science will likely conquer human mortality. After all, aging and biological death are just more puzzles for scientists and technologists to overcome. In the meantime, cryonics remains the best hope for transhumanists who desire to live indefinitely. Ensuring its complete legality is a top priority for them.

"Repeal the anti-cryonics law, already," says Ms. Gaspar. "We will not back down and we will challenge the legitimacy of this law."

*******

## 28) Should I Have Had My Cat Cryonically Preserved?

I recently made the agonizing decision to euthanize my cat Ollie, who I adopted 13 years before from the streets.

Ollie had barely eaten or drank anything for five days and was dying from kidney failure. The veterinarian told me Ollie would probably be dead in 24 hours and suggested euthanizing him, so that his death wasn't caused by choking or something horrible like that when other organs failed. I reluctantly agreed.

Pet euthanasia generally includes a heavy morphine-based sedative that peacefully knocks the animal out, followed by a heart stopper-chemical injection. We euthanized Ollie on his favorite couch in my home. While the process seemed painless and quick, it was absolutely heartbreaking for my family and me.

Days after the death, a number of transhumanist friends consoled me and told me of their own dealings with pet deaths. Since I'm a life extension advocate, I'm well-versed in procedures for dealing with (and avoiding) human death. But I didn't really know much about pets.

It turns out many transhumanists have already thought of these and some have even undergone cryonic procedures with their animals— the process where they cryogenically freeze their pets in hopes to resurrect them in the future when the technology becomes available. The Michigan-based Cryonics Institute has 120 frozen pets. Costs to freeze a cat there are about $5700 dollars.

I thought deeply about doing this with Ollie, but decided against it for a few reasons—reasons I hope I won't later regret in my life as the world and technology rapidly advances.

To begin with, I was a little late in the process with Ollie. It was already 24 hours after he died that I began considering cryonics for him, and, like humans, the cryonics process works best if it's begun within hours of death—especially to preserve the brain and its memories. Also, $5,700 is quite a chunk of cash—plus there are yearly maintenance costs. Additionally, my kids are already yearning for another pet, and my parents have had seven different pets so far in their lives.

With this mind, I even considered the cheaper preservation methods, where a bucket filled with formaldehyde, glutereldahyde, or some other solution is used to preserve the pet. Then one can just keep the body in their garage. In this procedure, at least much of the tissue, bones, and organs might be able to be salvaged in the future when trying to reanimate the animal. Some people even stuff their pet or freeze-dry them to keep them in their house, looking as if they were almost alive.

In the end, I passed on all these options and opted for a normal burial of Ollie in my backyard, which my young daughters and wife attended.

The truth is I tend to believe I'll be merged with AI in about 30-40 years—and soon entering the Singularity afterward—so the idea of loving a cat indefinitely seemed less tangible.

I also wondered if in the future, we'd be able—and maybe even obligated—to make our pets hyper-intelligent via cranial implant technology and radical genetics. Then the animal, like an adult offspring, becomes intelligent enough to make its own decisions. What if Ollie didn't want to live? Or be so intelligent? Or even be my pet anymore? Such is the weird world of transhumanist thinking— and the future many of us will face in the coming decades.

Either way, Ollie's death started me down exploring the road of technology and science we are going to impose on the creatures we love. It turns out the pet industry is exploding with transhuman—or if you will, transanimal—themes. Most of these have nothing to do with death, but instead have to do with giving animals a better life so humans can enjoy them more.

For starters, an entire cottage industry on pet-tech wearables has emerged, with numerous start-ups already competing in the space. *Vice Motherboard* reported there will likely be exhibition space specifically dedicated at CES 2017 to pet tech. Currently, the leading wearables are Fitbit-like devices that help monitor dogs' whereabouts and health.

Of course, pets have long had RFID chip implants to help locate them, and their success has led the way of implants into humans—such as the one I now have in my hand. But the future of tech for pets is also developing too. There are devices like TailTalk and the KYON collar that can supposedly tell you about your animal's mood. Some companies have even launched projects to try to directly read the brain waves of pets, so one day you might be able to discuss Plato's Allegory of a Cave—or the adventures of Garfield.

As cool as some of the tech coming out for pets is, the world is headed for a massive transformation about how and what it wants in its future pets. CRISPR gene editing is already here, and the idea of creating a pet dinosaur is no longer a pipe dream. In fact, MIT Technology Review reports that Chinese scientists have already created "designer pets."

It's possible in just a few years time we will be creating new creatures that contain the very best elements pets have. Shed-less dogs. Uber-cuddly cats. Melodic singing songbirds. Why not combine them? Why not add some reptilian genes too, for excitement? In fact, why not just make a make a mini-Brontosaurus?

Of course, the other type of future pet will be created by secretive company Magic Leap, where sensors on your ceiling can put out a holographic pet image that you can interact with and order around. Why not have an eight-foot tall Tyrannosaurs Rex inside to scare off burglars when they break in? Or a 30-foot anaconda? Or a pack of wolves? Best yet, you can program the holographic wolves to take turns reading your toddlers The Three Little Pigs.

The future of transhuman pets, though, is not holographic or biological. It's robotic. The field of robotic dogs already available on the market is massive. There are a few dozen companies and types of robotic dogs out there. Some of these machines are designed to be legitimate guard dogs, and can offer real security via movement tracking mechanisms and security software. In the near future, some will offer Skype abilities, so you can see through cameras in their eyes what's happening in your house—like if your child is playing with the stove. Other robot dogs will have built in fire alarms that can register smoke in a child's room or spot a poisonous spider in the dark crawling on a bed crib.

In probably just five years, robot dogs will be so sophisticated they will walk our children to school, carry our groceries for us from the car, and probably even have built in drone capabilities to fly. We'll program them to catch rats but not fight with the neighbor's cat. They won't need to be fed, they'll know how to recharge themselves, and gone forever will be the days of shoveling dog poo. And of course, they'll easily beat us in chess.

Some new pet robots have fake fur too. In the future, we can expect robotic pets to have non-shedding, clean smelling fur that is dirt resistant and looks just like a real pet. And the pet's bodies will be soft and padded, with heat creating capabilities to keep you warm at night when it sleeps with you.

Like so many other things technology is changing for the human race, the central role pets play in our lives will also change. The domestication of animals has evolved for thousands of years, but the next 25 years may end pet relationships as we know them. While I'm still a little unsure whether I should've cryo-preserved my cat, I think Ollie would've found it strange to be brought back to a world with chess-playing robot dogs, holographic wolves in the living room, and mini-Tyrannosaurus Rexs cruising around the backyard.

********

## 29) A Tragedy for a Young Cryonicist

Most people know of cryonics through Hollywood sci-fi movies such as Vanilla Sky and Demolition Man. Comedy films have also played with the concept, like Austin Powers: International Man of Mystery, where the film's arch enemies awaken from cyrostasis after a 30-year slumber and have comedic trouble adjusting to their new surroundings.

Cryonics works by preserving people's bodies in ultra-cold temperatures in hopes of reviving them in the future when better technology and medicine exist. Another far-cheaper method, costing generally just a few thousand dollars, is getting a high-concentration aldehyde brain preservation. Transhumanists hope that revived brains can be placed in robotic or genetically engineered human bodies in the future.

But society and its age-old beliefs of death are standing in the way of transhumanist goals. Many people—not to mention US state laws—don't support the rights of cryonicists.

A recent unfortunate case highlights this disconnect. 31-year-old Danielle Michelle Baker was found dead in the woods near her father's house in Kentucky in December last year. Baker was a cryonicist who attended life-extension conferences and had signed up to have her brain preserved via the nonprofit company Oregon Cryonics. She had both a cryonics contract—a signed legal document where she indicated that she would like her brain to be preserved for possible future life—and a Document of Gift, a legal document that allows her to donate her anatomical parts to a specific caretaker.

Upon hearing of her unexpected death in Kentucky, personal friends and Oregon Cryonics staff raced to present Baker's coroner and medical examiner the legal paperwork required to salvage her brain. Eric Homeyer, a volunteer representative of Oregon Cryonics, drove hours through the night from Cincinnati to reach the morgue to represent Baker's interests and make sure her body wasn't damaged or destroyed.

But despite the major parties knowing about the cryonics contract and Document of Gift, Baker's family pushed for the cremation, which then was carried out by the coroner via a funeral home three days later.

"This Kentucky coroner's office took it upon themselves to decide that Danielle would not get her final wishes fulfilled and would instead be cremated, blatantly and willfully breaking the law," Matthew Bryce Deutsch, one of Baker's friends, wrote on Facebook. While Deutsch is incorrect in saying that any laws were broken in Kentucky, cryonicists feel that violating the wishes of the deceased regarding their bodies should be against the law.

Cryonicists feel that violating the wishes of the deceased regarding their bodies should be against the law.

"The coroner always follows the wishes of the next of kin. It doesn't really matter what the decedent wanted," Jordan Sparks, executive director off Oregon Cryonics. "Most states do not recognize wishes of the decedent regarding the disposition of the body." In the US, only California, Arizona, Oregon, and Michigan have significant operating cryonics facilities where bodies can be preserved through an established legal process and procedure.

Because the great majority of US states don't abide by cryonics contracts, even if legally created by the deceased, cryonicists sometimes get around this by signing a Document of Gift, which essentially is a donor's card that allows you to donate parts of your body. This document is accepted in nearly all 50 states. But, as in Baker's case, it doesn't always work, especially if the recipient of the donation is not an accredited hospital or medical research center. (Most standard donations of human remains go to transplant patients or research centers.) Despite this gray area, cryonicists still usually sign both a cryonics contract and a Document of Gift, hoping one of the documents will be upheld to preserve them.

Some people consider cryonics quackery, and state that freezing the body's cells all but destroys the possibility of future life and retaining memories. But radical new experiments in recent years are changing our fundamental perspective of consciousness and death.

Last year, Yale scientists kept dead pigs brains alive outside of the body. Other researchers have had success implanting rodents with new memories created in a laboratory. Then there are some companies in the Bay Area, such as Elon Musk's Neuralink, that are even trying to create technologies that would allow us to upload our minds and thoughts to computers.

As life-preservation becomes a more mainstream alternative and more of these cases arise, we need to create a legal framework in which cryonicists' rights are protected. In response to what the cryonics community considers a tragedy, Deutsch has suggested the creation of "Danielle's Law," a legally binding framework of rules and guidelines that US states would implement to respect and honor human-body preservation contracts and Documents of Gift.

But as cryonicists are a tiny minority in the US and traditional views about death practices and rites persist, cryonicists are unlikely to have much sway in the near term unless science actually proves people can come back to life. In the meantime, dreams of coming back to life for people like Baker will go unrealized.

*******

# CHAPTER V: EXISTENTIAL RISK AND SURVIVAL

## 30) The Transhumanism Movement Aims to Eliminate Existential Risk for the World

Like many other people around the world, I am following the expanding Zika virus crisis. It's unsettling to watch images of people affected by the disease. Also, in the news, last week's earthquake in Taiwan tragically killed up to 100 people. And in the last few days, media is sensationally reporting a 30-meter wide asteroid soon to pass alarmingly close to the Earth.

These existential threat issues never seem to end, and they always make me ask: Does it have to be this way? Must we always worry about something? Or can science and technology eliminate all risks?

As a US Presidential candidate, it amazes me that more attention is not dedicated to overcoming existential risk by the very thing humans are good at: innovating. We can overcome all odds with new science and technology, and I personally advocate for spending much more government resources to do so.

Transhumanists are sometimes criticized as being overly optimistic about the future. It's true that transhumanist supporters do put nearly all of their hope into science and technology to solve the world's problems. However, last year's existential risk street action in San Diego—part of the Immortality Bus tour—showed a philosophically balanced approach to technology.

Transhumanist Party supporters made a public demonstration against existential risk in front of the San Diego Bay-anchored USS Midway aircraft carrier—now a popular tourist destination.

Carrying signs and banners bearing statements like "Make Love, Not Viruses", "AI Must Be Safe," and "Give NASA Funding for Asteroid Detection," nearly a dozen people gave speeches and engaged the public in discussion about planetary risks.

The street action was not the blind, rush-ahead syndrome that many luddites, religious media, and conspiracy theorists suggest all transhumanists possess with technology. Rather, transhumanists aim to use progress to eliminate harm and suffering, while also noting that too much technology—especially too fast—can be dangerous or harmful to the species.

Despite many countries around the world existing in relative peace and advancement, our species is under constant threat of cataclysmic existential risk—an event that might erase tens of thousands of years of human progress and literally send the species back into the dark ages.

Take Ebola for example. Ebola is a disease that can be conquered, and yet it ran havoc in America's psyche two years ago. In fact, since Ebola has been around for decades, the real question is: Why hasn't Ebola been stopped yet? Either with a cure or with a vaccine? Some experts believe Ebola could be mostly eradicated with less than a $100 million dollars, in the same way yellow fever, smallpox and other diseases have been mostly controlled. The net worth of the United States is about $125 trillion dollars. As a country, $100 million is a drop of water in the ocean for our citizen's health.

Unfortunately, the current US government—both Democrats and Republicans—don't consider such risks as necessary to tackle with the full power of our nation's resources and 21st Century science. I wonder if it'll take a Supervirus that kills tens of thousands to get the government to decide to spend its resources to eliminate major plagues.

Myself and other transhumanists are calling on politicians to realize that spending trillions of dollars on far-off wars, instead of spending a fraction of that money to potentially save our country from dire existential risks is plain irresponsible.

Recently, *The Boston Globe* reported that our government "is spending up to $1 trillion modernizing and revitalizing America's nuclear weapons."

Wow. $1 trillion. You could probably wipe out nearly every single major disease on the planet with that kind of money. But more importantly, as a species, do we really need 25,000 nuclear weapons on the planet—because that's the amount that's out there? Between possible accidents, terrorism, and warmongering dictators, something bad easily could happen from them. I say let's get rid of all nukes before some tragedy strikes.

There's something very wrong with a society that makes so much of its GDP from a military industrial complex bent on war. We should aim to create a replacement for it that focuses on science. And among other priorities, that science industrial complex should protect us from existential risk.

In my opinion, the government is not distributing the people's resources properly or in their best interest. In the future, I hope politicians in America will deal with existential threats far ahead of time, instead of leading us down a path where something truly awful might happen to Earth

********

## 31) We Must Destroy Nukes Before an Artificial Intelligence Learns to Use Them

On a warm October day, videographer Roen Horn, *Slate* columnist Mark O'Connell, and I visited the White Sands Missile Range, which is home to an outdoor missile museum. Still an active military base, it takes a 10-minute security inspection and a stamped clearance to even be able to enter the area.

I was in the middle of political campaign, and we were after a photo-op of transhumanists in front of missiles. The fact that 25,000 nuclear bombs exist around the world is reason enough to try to warn the public. A detonated bomb in a mega-city could easily kill hundreds of thousands.

Generally speaking, many transhumanists strongly advocate for using science and technology to eliminate death. This means much of the movement and its scientists are focusing on stopping aging through gene therapies, overcoming disease through robotic organs, and living longer via radical diets or imbibing a handful of pills every day.

While I actively support all these longevity techniques, it's important to realize that death doesn't just come through the human aging and disease, but also through human folly. This is why I believe transhumanists should also be deeply concerned about existential risk, such as nuclear warfare. I personally advocate for dismantling all nuclear weapons.

Thankfully, it's not just transhumanists who are concerned about such risks. Plenty of politicians in recent campaigns noted that nuclear threat is a great worry.

To me, it's critical that major politicians continue to be vocal about this ongoing dilemma. As a society, we have become complacent to the fact that so many nuclear arms are still out there. But tensions between Russia and America, as well as those of China and America, are a stark reminder we should not ever get complacent. The ongoing rivalry between India and Pakistan, also nuclear heavyweights, should be considered too.

The first nuclear bomb, called Little Boy, was dropped on Hiroshima, Japan on August 6, 1945. Eighty thousand people immediately perished.

Six years before, the creation of the bomb began with a suggestion by Hungarian physicist Leo Szilard and Albert Einstein to President Franklin D. Roosevelt, which culminated with the Manhattan Project, perhaps the most impressive of any technological effort the United States has ever engaged in.

It's sad to me that one of America's greatest technology efforts is one of war. However, it's not only open war civilization should be worried about; it's thievery or malfunction of nuclear arms, too.

Additionally, a major coming worry about nuclear weaponry is the rise of artificial intelligence and a Terminator scenario unfolding. I'm quite certain the highest long term military priority of America is the development and containment of coming artificial intelligence, because whichever country creates a superintelligent AI first would probably have the ability to break and rewrite all nuclear codes on the planet. Such power could change the global political landscape overnight.

Walking around the missile park—with its big yellow signs warning visitors to beware of rattlesnakes—it's impossible not to get the feeling this is war exhibit dedicated to men and their toys of fighting. An indoor museum is attached to the outdoor missile park, and nearly every portrait of the missile base founders and operators was of a gray-haired Caucasian man steeped in the means of warfare. Looking at those pictures, I longed for a new, diverse generation of humanitarian-minded influencers and leaders to come reinvent how we handle the security of Americans and people on Earth. Clearly, there must be a better way than the ironic amassing of tens-of-thousands of four-story tall killing machines.

*******

## 32) Genetic Editing Could Cause World War III

While *Time* magazine recently chose President-Elect Donald Trump as its Person of the Year, CRISPR gene editing pioneers were a runner-up choice. Few innovations in the last millennium carry such transformative prospects as the ability to edit our own genome and make ourselves into fundamentally something else. Some experts think genetic editing might be the key to curing all disease and achieving perfect health.

Unlike other epic scientific advances—like the 1945 explosion of the first atomic bomb in New Mexico—the immediate effect of genetic editing technology is not dangerous. Yet, it stands to be just as divisive to humans as the 70-year proliferation of nuclear weaponry. On one hand, you have secular-minded China and its scientists leading the gene editing revolution, openly modifying the human genome in hopes of improving the human being. On the other hand, you have a broadly Republican US administration and Congress that appears to be strongly Christian—conservatives who often insist humans should remain just as God created them.

Therein lies a great coming conflict, one that I'm sure will lead to street protests, riots, and civil strife—the kind described explicitly in my novel *The Transhumanist Wager*, where a religious-fundamentalist government shuts down extreme science in the name of conservatism. The playing field of geopolitics is pretty simple: If China or another country vows to increase its children's intelligence via genetic editing (which I estimate they will be able to do in 6-12 years time), and America chooses to remain "au naturel" because they insist that's how God made them, a conflict species-deep will quickly arise. If this scenario seems too bizarre to happen, just consider the Russian Olympic track and field team that was banned in the recent 2016 Games for supposed doping.

It's quite possible the same accusatory flavor of "banning" could happen between China and America in the game of life—between its workers, its politicians, is people, its artists, and its media. I wonder if America—approximately 70 percent who identify as Christians—will put up with beings who modified themselves by science to be smarter and more functional entities.

This type of idea takes racism and immigration to a whole new level. Will America close off its borders, its jobs, its schools, and its general openness to the world to stay pure, old-fashioned human? Will we stop trading, befriending, and even starting families with those who are modified?

In short, will genetic editing start a new cold war? One that bears much finger pointing and verbal reprimands, including the use of derogatory terms like mutants, cyborgs, and transhumanists. Think the videogame Dues Ex, but with modified people taking all the best jobs. In a worst-case scenario, it could even start a World War.

So, now that we know what can happen if America won't embrace the most important science to emerge this century, how can we avoid it?

First—and this is wishful thinking, since 100 percent of the US Congress and the Supreme Court appear to be religious at the moment—is we could just embrace genetic editing and be better at it than the Chinese. This is the exact scenario I suggest. Yes, it will lead to a place where beings are similar to those in Star Wars and Star Trek, but after all, we love those stories because we want to reach that super-science age. And in the long run, such evolution of the species is inevitable anyway, so long as we don't kill ourselves first in a nuclear war or an environmental catastrophe.

In a second scenario, America could focus more on technology and less on biology and genetics. On my recent 4-month long Immortality Bus tour across America, I found conservative people seem more inclined to use tech accessories or wear a special headset that would make them smarter (for example, by connecting their thoughts Matrix-style into the cloud and AI)—as opposed to structurally changing their brains, as the Chinese likely will do. America could innovate that accessory tech that would keep us ahead of the biological modifications of other nations. I'll accept that—reluctantly—if the first scenario I presented is a no-go.

A third way—and this is the blatant transhumanist nightmare—is we could establish a non-modification policy across all countries, similar to how we have created the Paris Treaty for climate change or rules of war that ban chemical weapons. The major nations of the world, sensing a significant global legal issue in genetic editing, could come together as a species and criminalize the science.

To some extent, this has already happened, because as soon as the world realized the Chinese had experimented on the human genome, calls were made to put a stop on some of this science. Such a reaction is not dissimilar from what George W. Bush did with stem cells when his religious values made him shut down federal funding on all but a tiny portion of the research in America. Stem cells have since been shown to be one of the most important medical applications in the world, and those lost years of science have potentially negatively affected millions of lives.

Sadly, the third option of a general or even partial moratorium on genetic editing will surely harm innovation. The great thing with gene editing is we can likely do many wondrous things with it, such as potentially cure cancer, halt aging, grow better organs, and overcome disability by better repairing ourselves. Beyond making ourselves superhuman, we can simply make ourselves better fit for Earth, including dealing with a changing environment.

I also don't think the third option will work in the long run. More than ever, science is the hands of individuals, who can buy amazing biotesting kits on eBay for just a $1000—as well as incredibly powerful computers to analyze the data. Citizen scientists would just create the new gene editing tech and begin doing it themselves—perhaps more dangerously had the government not been overseeing the research from the start.

I argue for the first path. Let's allow good, old-fashioned scientific competition with China to proceed. Let's see which country can create the best enhancements for their citizenry, and let's share the best of our work with one another in the end to make it so all peoples are as equal as possible. If we're too closed-minded about such radical science, we might find ourselves embroiled in a state of hostile speciation, where another new cold war—or worse—swallows a generation.

*******

## 33) Four Technologies That Could Let Humans Survive Environmental Disaster

Scientists say we blew it. We bought too many plastic trinkets from Walmart; we drove too many gas-guzzling Broncos. We made babies like rabbits without questioning if the planet could handle so many people. Well, it looks like it couldn't. Climate change is here to stay, and it'll probably end up affecting nearly every aspect of our lives over the next century.

Like you, I'm not happy about this. But there are potential solutions. Not the ones politicians and environmentalists are pushing, like recycling, driving electric cars, and lessening our carbon footprint. These are things I support, but I believe it's too late to stop climate change. The way I see it, the race we're in now—the challenge of the century for our species—is how quickly humans can adapt beyond our biological selves versus how quickly we destroy the planet.

Our species' fate could hang on our transhuman evolution into a cyborg or machine-like state that's far less vulnerable to environmental conditions.

There are four critical technologies humans will need to survive most environmental catastrophes—be it the changing climate, a large asteroid hitting the planet, or nuclear war (in case you forgot, the world still has about 25,000 nuclear weapons).

## Immunotherapy

The first is new treatments for cancer caused by ozone layer depletion, increased petrochemicals in food production, or radiation fallout from disasters like Japan's Fukushima meltdown.

Oncology has always hoped for a one-time cure for cancer, but the disease is so complex and varied that there may be no single magic bullet to be found. However, the field of immune-oncology—using the body's own defense system to attack cancerous cells—is emerging as a promising new way to successfully treat the disease. It could end up rendering cancer far less deadly.

"Within 10 years almost everyone suffering from cancer will be treated with an immunotherapy, Dr Hoos, Vice President of Oncology Research at GlaxoSmithKline recently told Australia's Herald Sun. One of the reasons many doctors are jumping on the immunotherapy bandwagon is that it may eliminate the need for toxic chemotherapy as a treatment. Instead, immunotherapy combines a number of other treatments based on the body's natural defense system to make cancer less disruptive to patients' lives.

## Bionic Organs

Many people die because of organ failure, which is caused by many things, including cellular death, disease, and trauma. One major key to overcoming environmental threats to our biological bodies is better, more durable organs, including robotic ones. A revolution in this field is also underway.

French biotech company Carmat is already installing artificial hearts (pictured left) in patients that can last five years. There's a chance these bionic hearts will one day be the equivalent of human hearts in how effective they are at moving blood (which carries all-important oxygen) around the body.

Meanwhile, we are making progress 3D printing organs, such as livers. While we have yet to produce a fully functioning organ, researchers have delivered created liver tissue samples outside the lab. These pieces of organ tissue provide way for researchers to test radical new drugs on organ cells without endangering a live patient. Scientists have already recreated skull bones, noses, and arteries, and other human body parts using 3D printed biomaterials.

But perhaps the greatest need in any environmental catastrophe is clean air to breath. The fact we need oxygen at every moment of our lives to operate normally is human beings' major biological weakness—perhaps even more than eating. Artificial or bionic lungs could be an alternative.

Many hospitals have ventilator systems that assist nonfunctioning lungs and allow patients to survive when they can't breathe properly, but so far they are complicated and bulky machines. Humans need such respiratory systems built inside themselves. BioLung is an option, an artificial device the size of a soda can that tries to replicate a lung. Human trials will start within two years.

According to WebMD, "University of Texas researcher Joseph Zwischenberger, MD, tried out the BioLung on sheep whose lungs had been badly burned by inhaling smoke. Six of the eight sheep on the BioLung survived five days, doing far better than ventilation machines."

Another option is "oxygen shots," a new technology already on its way. The technique involves injecting oxygen-filled microparticles into the bloodstream. Imagine it as something you might take a few times a day when breathing is difficult or impossible.


*Wearable Tech for Temperature Control*

Of course, it's not just about staying alive. We want to thrive, not just survive. Which leads me to a third technology we absolutely would need: sophisticated wearable tech that keeps people cool or warm in the harshest conditions, whether it's global warming or an ice age.

Our skin and temperature regulation can handle extreme temperatures for short periods of time, but long term, we need a stable, suitable climate. Some developing wearable tech is going this direction, offering cooling or heating built into the material.

Fuel Wear Flame Base, a crowdfunded smart-heated base layer, is meant to protect against extreme cold. It electronically warms up your clothes to keep you at just the right temperature, which could be really useful if the planet goes into a mini-ice age. Other clothing like those from ClimaWear (pictured left) claims to keep you cool when it's super hot—just what we need when the polar ice caps melt and the world warms 20 extra degrees.

Indeed, ideally, the more cyborg parts we get, the less that heating or cooling will be an issue. Some biohackers hope to electively replace limbs with bionic ones within 10 years. These robotic limbs can already tie into our neural system and react to our thoughts, and they won't need heating or cooling, lessening the overall burden on the body.

*Food Alternatives*

To speculate even further, a more mechanized body like that of a robot would not need food or water, that other great example of humanity's fragility. But what if we could eliminate eating and drinking altogether?

Frankly, I think doing away permanently with caloric intake should probably be a goal of all humanity, since it presents massive hassle from an evolutionary point of view. Millions of people die worldwide every year from food poisoning, digestive issues, and especially malnutrition. The fewer organs we need (including our bowels) the better off we'll be, and not needing food or water would dramatically lesson our reliance on the environment. If we were mostly integrated with machines, we could instead use other energy sources to power our bodies, like solar or fusion.

But let's come back down to Earth for now. If we can't (or don't want to) eliminate our dependence on food, then it would be wise to find an abundant alternative calorie source that is healthy and nutritious. Something like Soylent, which is healthier than many corn-based products, and in the long run will not require as much food production to feed the world, especially livestock production.

Another food substitution might be synthetic food—edible, energy-providing substances created in a laboratory. The most promising is synthetic meat grown from stem cells. Meat is important since many people around the world eat it, and it's dense in nutrition and calories. Creating cheap synthetic meat could mean much less reliance on Earth's resources, since currently 30 percent of the Earth's usable land is dedicated to livestock and animal grazing.

Even in the next 50 years, life could get wild and dystopic. But I believe that no matter how bad the planet gets, our species can still thrive, by becoming a species not so dependent on Mother Earth, but rather on the technology we can create. It's very possible we'll be able to successfully do that, so long as we're realistic with ourselves.

The strange truth is that in the very near future we may not need clean water or stable weather to sustain advances in our species. We may not need oceans or forests to feed us, or an ozone layer to help us avoid cancer. We may not even need the sun to shine bright, if we can perfect other forms of power. Existential risk can be overcome by adapting beyond the fragility of human biology.

********

## 34) Space Exploration will Spur Transhumanism and Mitigate Existential Risk

When people think about rocket ships and space exploration, they often imagine traveling across the Milky Way, landing on mysterious planets and even meeting alien life forms.

In reality, humans' drive to get off Planet Earth has led to tremendous technological advances in our mundane daily lives — ones we use right here at home on terra firma.

I recently walked through Boston's Logan International Airport; a NASA display reminded me that GPS navigation, anti-icing systems, memory foam and LED lights were all originally created for space travel. Other inventions NASA science has created include the pacemaker, scratch-resistant lenses and the solar panel.

These types of advancements are one of the most important reasons I am hoping our next U.S. president will try to jump-start the American space program — both privately and publicly. Unfortunately, it doesn't appear any of them are talking about the issue in a serious way. But they should be. As we enter the transhumanist age — the era of bionic limbs, brain implants and artificial intelligence — space exploration might once again dramatically lead us forward in discovering the most our species can become.

Already SpaceX, led by CEO Elon Musk, has announced it will be tackling an unmanned trip to Mars in the near future. The hope, of course, is that within the next 10 years, astronauts will be stepping foot upon the red planet, too. If indeed, humans can make it to Mars — and I'm sure we will — much new tech would have to be developed for the mission. It's safe to say much of that tech would likely be something useful for us eventually on Earth, as well.

For example, just to even live in space for the journey — it'll take approximately six months to travel one-way to Mars — new ways of sleeping, recycling breathable air and preserving foods and drink would likely have to be developed.

Furthermore, the technology to withstand massive dust storms, freezing temperatures and a hostile environment on Mars would require new space suits and maybe even totally new materials. Innovation like this will benefit everyone — even if we don't know all the uses yet for such radical tech.

Of course, there are other reasons for prompting a renewed and significantly larger space program in America. One of the fundamental goals of my own presidential campaign has been warning the world of the incredible threat of existential risk.

*The Atlantic* recently ran a story by Robinson Meyer that read:

*At life-long scales, one in 120 Americans die in an accident. The risk of human extinction due to climate change — or an accidental nuclear war — is much higher than that. The Stern Review, the U.K. government's premier report on the economics of climate change, estimated a 0.1 percent risk of human extinction every year. That may sound low, but it also adds up when extrapolated to century-scale. Across 100 years, that figure would entail a 9.5 percent chance of human extinction.*

I think most people are totally unaware at how high the odds are that we screw up our species' very existence. It's so high, that the newly written *Transhumanist Bill of Rights* has a mandate for space exploration as one of its key six points.

The facts of existential risk are simple: We may not be able to indefinitely keep the planet habitable, stop a super virus from killing everyone, avoid a mile-wide asteroid from crashing into Earth, elude a warmongering Terminator-like AI or circumvent blowing ourselves up with our 25,000 nuclear warheads — but we sure can get off this planet and create cool new places to live safely in outer space.

The movie Elysium recently showed a dystopic but technologically plausible space habitat, where paradise is engineered in the skies — and not on Earth's land or water. Now, no one wants to be forced into this scenario, but massive space habitats are worthwhile projects to pursue — and they could be possible to build in as little as 15 years.

Mega-space habitats would also make an easier launch base for space mining, an industry booming with interest. Experts say it will soon be possible to mine asteroids from space — some that are worth billions of dollars each. Clive Thomson at *Wired* recently wrote that the asteroid Ryugu — partially made of up of nickel, iron and cobalt — could be worth up to $95 billion.

As a science advocate, I'm strongly pro-space exploration from a private industry point of view. But just as importantly, I also passionately support a U.S. government-sponsored space program — one that gets approximately 10 times the funding it gets now (I'd get that extra money from our military budget, which is oversized anyway). That would be nearly $100 billion a year, or about 5 percent of the U.S. 2016 Federal budget.

Generally, my fiscally-minded self doesn't want the government too involved in much of anything, but because space exploration involves defending against existential risk and pursuing medical innovation for citizens, I'd advocate for the U.S. putting dramatically more resources into space exploration. This wouldn't mean entirely relying on federal programs to push forward the space industry, but also on government partnering with or investing in private space companies.

Sadly, Congress will likely put up a fight against spending too much on peacetime space exploration — they do have that habit of being boring and shortsighted. So, perhaps the best way to grow America's space industry is to sell Congress on the amount of benefits our nation might gain from a meaningful and dramatically enlarged space program. Generally, politicians — those directly responsible for funding (or not funding) NASA — see no upside for sending astronauts to space except national pride.

But if Congress could be convinced that national security against existential risk, money from space mining, and precious tech innovation for U.S. citizens would be gained by supporting space exploration, then maybe they would vote to enlarge NASA's programs. This in turn would spur both the private space industry as well as transhumanism tech that makes all our lives better. This type of thinking should be a priority for whoever ends up in Congress and the White House this next election.

*******

# CHAPTER VI: SOCIAL ASPECTS OF LIFE EXTENSION

## 35) Marriage Won't Make Sense When We Live 1000 Years

I was jubilant the US Supreme Court ruled in favor of gay marriage. Events that lead to more freedom and equality are positive progress.

However, what doesn't seem to be making the news is the fact that marriage—especially to many young people—isn't as attractive as it once was.

There are a number of reasons for this. People want to focus on their careers, not spouses. Getting married and having a traditional wedding costs a lot of money (besides, around 40 percent of those who wed will go through at least one divorce in their lives, causing potential harm to their ideals, children, and finances). Finally, having kids out of wedlock is becoming more socially acceptable.

But there's another reason that is increasingly relevant. It has to do with transhumanism. In the transhumanist age of extended lifespans, where many people will live beyond 100 years of age, the question of being married until "death does us part" has real consequence.

In America most marriages last about a decade. However, it's safe to say that plenty of those marriages that do last much longer are not entirely happy or fulfilling. Fear of being alone, apathy, and finances often bind the reluctant wedlock yoke. But I believe the primary reason people stay married when they're not happy is religion. Some Abrahamic religions treat divorce as sin (thereby potentially jeopardizing one's afterlife if you get divorced). Especially in America where some 80 percent of the citizenry is Christian, faith plays an influential part in promoting marital union.

Social, financial, and religions pressures aside, the deeper philosophical question of the transhumanist age is: Are people really willing to marry for the rest of their lives when those lives may be hundreds or even thousands of years long? This is especially a pertinent question when it's almost certain coming technology will allow us to radically change who we are in the near future, both physically and mentally.

In a world of indefinite lifespans, the marriage commitment takes on a whole new meaning and level of commitment.

America and many parts of the developed world are losing their religion, however, which certainly will contribute to less social pushing for matrimony. A recent Pew Research Center study found that many young people increasingly possess no religious leanings at all. In just a few decade's time, if this statistical trajectory holds, younger generations may broadly prefer not to ever marry.

And who can argue with them? Within 15 years, some of the so-called classic advantages of marriage will be gone. Many people will have robot house nannies, driverless cars, and automated stoves that cook for us. In 20 years' time, we may also use artificial wombs (ectogenesis) to grow babies, and use our own stem cells to provide genetic treatments to build the perfect child. A spouse will simply not be as necessary in the transhumanist age as it once was.

Naysayers will argue that only a wholesome, traditional family can produce good, well-rounded children. But that's deeply wrong. In 15 to 20 years' time, cranial implant technology will enable humans to overcome many of their idiosyncrasies and bad behaviors—making a new generation of very wholesome and exemplary children. In fact, going to college may be replaced by downloading higher educations into our brains.

Even morality may be built in by personal avatars that are always looking over our shoulder for us, not dissimilar to what Abraham Lincoln called the "better angels of our nature." In just over a decade, traditional family life and the institution of marriage as we know it will face the largest disruption it's ever gone through.

And sex? Well, that can and will be better and more pleasurable with the rise of transhumanist technology. Already, scientists are working on pure, outright stimulation of the erogenous zones in our brains. Stimulating this part of ourselves will be easier, on-demand, and disease and pregnancy-free. Of course, the coming world of virtual and augmented reality will also offer endless amounts of physical experimentation via haptic suits to satisfy one's lusts, too.

Another thing sure to make people—both young and old—wary of marriage in the future is the growing promise of gender-identity choice. In the transhumanist age, we are not stuck being males or females, but whatever version we want—maybe even something between or combined. Transgender surgery is catching on and people can change themselves as they see fit, or they can do it just for kicks and new experiences. In fact, most of the modern medicine, surgery techniques, and tech are already here today or coming soon—complete with augmented penises, vaginas, and other sexual body parts that we can replace or modify.

But the bigger transhumanist steps of gender and identity will come when we begin uploading our minds into machines, and people must decide what their avatars will be like. Surely, many people will experiment with other sides of themselves they always wondered about. Think of uploading as an anonymous masquerade party, where you can be anything you want, and then be something else later that day. People may change their genders daily, depending on who they interact with or how they feel.

All this radical tech and change the human race is about to undergo means one thing: marriage is heading the way of the dinosaurs. So instead of celebrating our rights of matrimony for gay people or trying to privatize it for tax and liberty reasons, maybe we should also begin endorsing the phasing out of marriage from society's mindset.

Of course, that doesn't mean we won't have intimate relationships that are deep and meaningful. It just means that the multi-millennial-old institution of marriage—began by our ancestors to transfer inheritance in the form of dowries (often weapons and livestock)—has increasingly less relevance today. In the meantime, we'll come up with new ways to create legal structures to protect relationships and those we love in a deeply litigious society. In nearly every instance of legal companionship, a simple notarized document giving permission to a partner can serve where a marriage certificate once did the same. In the future, this legal procedure won't be physical, anyway, but notaries and permissions will be done by large database scans of retinas, fingerprints, and DNA samples on your smartphone or chip implants.

Even though I'm a happily married man with two kids, I'm all too aware of how society, the government, and especially religion has sold people on the concept that love needs to be institutionalized and consummated by legal marital vows.

In my opinion, that's all just another level of control someone or some entity is trying to put over me and others. If one is in love, then they need no controls. Love just is, and for two people in love, it manifests itself every day. And if it doesn't, then it's no longer love. Society can operate on a new social structure that incorporates other versions of social bonding, ones that also support a strong, caring, and connected society. This includes stepping away from all-holy monogamy, and implementing a larger mindset about what constitutes relationships.

For the record, I'm not saying let's throw away marriage. But let's stop society and government from promoting it like it's the only way to love and exist.

In the transhumanist age, it's time to leave behind closed-mindedness. In our relationships with others, we should instead look not with our biases and bigotry, but for what a person we care about can do for us, and what we can do for them. That person may be a human, a cyborg, a robot, or even a computer program. Whatever it is, frankly, is not important. It's what it does and how it does it. And if it does good, honest, and meaningful actions, then that's plenty upon which to build love, intimacy, and a successful future.

In fact, soon, the next civil rights debate of love and marriage will probably involve whether we can wed the coming generation of intelligent robots and avatars, which may be nearly as smart as us in a decade's time. This brings up larger questions of different legalities. It also brings up polygamy. Is being wed to two robots at the same time more socially acceptable then marrying two human spouses? Will the US government support tax breaks of marrying robots as it does for humans (as President, I would advocate for this)? Will divorce laws be different for the machines we wed— assuming they'll agree to wed us at all. Will divorces be governed by communal law or common law? Do we need consent to marry a machine? We surely don't need any to fall in love with one.

The coming transhumanist age is indubitably thorny. The onslaught of radically technology in our lives is challenging the very institutions and ideas we have built society upon. However, I hold much hope that technology will continue to allow us to live longer, better, and freer. Whatever happens, we shouldn't remain mired in past practices that once served society, but no longer do in such a positive or functional manner. We must look forward and search out new ways of living that grant us improved livelihoods.

********

### 36) I Visited a Church that Wants to Conquer Death

Many people think of transhumanism — the belief that humans can evolve through science and tech — as a secular movement. For the most part it is, but there are a number of organizations that aim to combine science and spirituality together.

One of the largest is the Church of Perpetual Life, a brick and mortar worship center near Miami, Florida that looks like any other church. It has a minister, a congregation, and church activities. The only difference is this church wants to use science to conquer death.

I was asked to speak at a Church of Perpetual Life service while traveling across America on my Immortality Bus — a coffin-like campaign bus I'm using during my run for president of the US (under the guise of the Transhumanist Part). Services at the Church of Perpetual Life don't revolve around worshiping a deity. They're passionate exploration of life extension research. It's a group of people that want to live forever, but also want belong to a spiritual community.

Conversations are centered around how humanity can improve itself through science, how we can overcome death with technology, and how suffering can be broadly eliminated.

The church itself welcomes people of all religions, and sometimes explores concepts of a benign creator in very nonspecific terms. But mostly church services are dedicated to hosting invited speakers who make presentations on the current status of the anti-aging field. For example, entrepreneur Martine Rothblatt of Terasem recently spoke here.

The Church of Perpetual Life—whose symbol is a fiery phoenix— was originally founded by multi-millionaire Bill Faloon and his business partner Saul Kent. Faloon is known widely in the transhumanist community for being a cryonicist, and he has helped fund many life extension projects.

A cornerstone of the philosophy of the Church of Perpetual Life is its interest in the 19th century Russian prophet Nikolai Fyodorovish Fyororov, considered by some an early transhumanist. He believed we could follow science to become our best selves, and that is was the task of humanity to conquer death and unite humans in love and peace. The Church of Perpetual Life considers him a prophet.

Major church services take place about every month, and sometimes more frequently if a longevity speaker is in town. Neal VenDerRee, the certified minister of the church, presides over the sermons. He is also the main go-to person of the 500+ person congregation. VanDerRee and I spoke a number of times on Buddhist philosophy, which both of us appreciate greatly.

On the night I spoke, the 38-foot-long Immortality Bus was parked by the church entrance, with flood lights hovering over it. Because the bus resembles a giant coffin (to remind people we should all be working on overcoming death), the church decided to put a spotlight on it after the sermon for the 60 or so churchgoers.

Because I'm a US presidential candidate, my speeches are almost always political. But I promised VanDerRee I wouldn't speak at all about politics. So instead, I spoke on how important it is spread transhumanism to the general public. There were discussions about the implant I have in my hand, stem cell technology, and whether mind uploading is possible. I believe it is.

After my speech, Bill Faloon gave a short passionate talk on the dangers of high blood pressure, and transhumanist Maitreya One performed a short rap song about longevity. The evening ended with drinks and dinner, as well as visits aboard the Immortality Bus.

Leaving the Church of Perpetual Life made me think about my atheism. After being raised a Catholic, and even attending Catholic school where religious dogma was drilled into me, it was refreshing to be inside a church and feel part of a spiritual community without all the threats of damnation.

A church that asks nothing from you and hopes to end death for all humanity using science—now that's something I can support.

*******

## 37) Oligarch Pledges $1 Million Prize to the First Person that Can Live to be 123

A Moldovan multi-millionaire whose dream it is to live forever has promised to give $1 million to the first person to reach the age of 123.

Dmitry Kaminskiy, a senior partner of Hong Kong-based firm, Deep Knowledge Ventures, is hoping his million dollar gift will trigger a new group of 'supercenternarians'.

He says research into stem cells, tissue rejuvenation and regenerative medicine will allow people to live beyond 120 - an age that has been quoted as the 'real absolute limit to human lifespan'.

'We live in the most exciting era of human development when technologies become exponential and transformative,' Kaminskiy told *DailyMail.com*.

'They may not realise it, but some of the supercentenarians alive today may see the dawn of the next century if they live long enough for these transformative technologies to develop.

'I hope that my prize will help some of them desire longer lifespans and make their approaches to living longer a little more competitive.'

French born Jeanne Calment currently holds the record, having lived to age 122.5. She died in 1997.

Already, a number of supercentenarians are candidates to best Calment's record.

The oldest verifiable supercentenarians living now are Jeralean Talley, at 115 years, and Susannah Mushatt Jones, also at 115 years. Both are Americans, and Talley is older by 44 days.

In the last few years, major anti-aging companies, such as Google's Calico and J. Craig Venture's new San Diego-based genome sequencing start-up Human Longevity Inc, have launched.

Along side this, Billionaires like Larry Ellison, Sergey Brin, Peter Thiel, Paul F. Glenn, and Dmitry Itskov are also funding research into longevity science.

Itskov is the founder of the 2045 Initiative with the goal of helping humans achieve physical immortality within the next three decades.

Reuters reported that gerontologist Dr Aubrey de Grey, chief scientist at SENS Research Foundation and Anti-Aging Advisor to the US Transhumanist Party, thinks scientists may be able to control aging in the near future.

'I'd say we have a 50/50 chance of bringing aging under what I'd call a decisive level of medical control within the next 25 years or so,' he said.

Kaminskiy is hopeful people will soon start living to 150 years of age with longevity science improving.

Studies in stopping and reversing aging in mice have already shown some success, and people around the world are generally living longer all the time.

For example, life expectancy hit an all time high of nearly 79 years-old in 2014 in America according to a report from Centers for Disease Control and Prevention's National Center for Health Statistics.

Additionally, Britain's oldest person, Ethel Lang, just died at the age of 114. She was believed to be the last person living in the UK who was born in the reign of Queen Victoria.

Whether Kaminskiy's million dollar prize will encourage people to live longer is still to be seen.

But this isn't the first time Kaminskiy has made news with in the longevity field using his resources.

At January's JPMorgan Health Care Conference in San Francisco, he recently bet Dr. Alex Zhavoronkov, PhD, CEO of anti-aging company Insilico Medicine Inc. for a million dollars in stock who would live beyond 100 years of age.

'Longevity competitions may be a great way to combat both psychological and biological aging,' Dr. Zhavoronkov said.

'I hope that we will start a trend.'

Zhavorokov thinks longevity science today resembles the computer industry in the 70s or networking in mid-90s.

'Most of the pieces are there,' he said. 'We just need a product or a service to transform the way we live. The revolution in longevity is just around the corner and it is time to seriously engage in the field.'

Historically speaking, prizes have made a difference in the way science and culture has evolved, and also in the way people look at the world.

The Nobel Prizes, given out in Stockholm, Sweden every year, are perhaps the most famous and coveted of all awards on the planet.

Scientists and peace activists careers can change overnight by winning the prestigious prize.

It's possible the million dollar prize for the longest living person ever on the planet will also evolve into a well-known award.

Perhaps other wealthy enthusiasts will step forward to offer new longevity awards after a supercentenarian reaches age 123.

Maybe a trend will be set, and new awards for age 135, or even age 150 will be established.

## 38) For Christians, Does Being Pro-Life Lead More Souls to Hell?

In late November, the Colorado Planned Parenthood shooting where three people were killed and nine wounded sadly reminded Americans again that women are not safe in this nation when trying to make choices about their bodies. It compelled me to take a candid spiritual look at the popular Christian stance on abortion.

As a transhumanist US Presidential candidate, I am pro-choice. More interestingly, though, as a former Christian and Catholic school student, I was also pro-choice.

Here's why: From a strictly biblical point of view, being born on Earth is a test. All our actions will eventually be judged by an omnipotent God who will determine whether we go to heaven or hell. Our deeds, sinful or not, determine where we spend eternity. The Bible even says many people will not get into heaven because it's quite difficult to be a sin-free Christian—meaning the majority of human souls may spend an endless amount of days in hell. Like it or not, about 2.2 billion people on Earth believe in these ideas. Another 1.6 billion Muslims believe in mostly the same thing, too.

Where the Abrahamic rabbit hole gets weird—at least for me—is the fact that many Christians (and Muslims) believe that an aborted fetus goes to heaven.

The metaphysical impact of that religious belief is just bizarre. It means that the most sure thing to do to get a soul into heaven is to abort a fetus before it leaves its mother's womb and has the chance to sin. The crazy thing is this makes abortion providers some of the most considerate, humanitarian people we know—at least from an Abrahamic religious perspective. Abortion providers and pro-choice advocates have long been filling heaven with pure souls—instead of committing them to a lifetime on Earth, challenged with trials of sex, drugs, and transhumanism.

Of course, the logic described above certainly sours the pro-life argument that abortion is evil. However, the premise of that logic is anything but certain or sound. It requires basing views on leaps of faith and ancient religious mores—like the existence of God, hell, and heaven. The reality—despite the billions who believe in formal religion—is that no one really knows anything definitely. While I like to say I'm atheist just to be defiant in the face of an overly religious civilization, like any sensible person, I honestly don't know what exists beyond me and the material universe I live in.

The fact is—since we're all empathetic mammals—that no one likes to have abortions, and no one likes to provide them. In an ideal world, no one would ever get pregnant unless they were certain they were ready for parenthood and knew they were going to have a perfect child. But stuff happens and things don't go as planned, and people must do things to try to accomplish the greater good for themselves and society. In this way, we should be grateful that women's clinics—like Planned Parenthood—are there to help out so that the best choice about one's life can be made.

And if we insist on being a Christian, then we might want to look at the bigger spiritual picture and stop trying to shoot people that are helping a soul's entrance into heaven.

**********

## 39) Environmentalists are Wrong: Nature Isn't Sacred and We Should Replace It

On a warming planet bearing scars of significant environmental destruction, you'd think one of the 21st Century's most notable emerging social groups—transhumanists—would be concerned. Many are not. Transhumanists first and foremost want to live indefinitely, and they are outraged at the fact their bodies age and are destined to die. They blame their biological nature, and dream of a day when DNA is replaced with silicon and data.

Their enmity of biology goes further than just their bodies. They see Mother Earth as a hostile space where every living creature—be it a tree, insect, mammal, or virus—is out for itself. Everything is part of the food chain, and subject to natural law: consumption by violent murder in the preponderance of cases. Life is vicious. It makes me think of pet dogs and cats, and how it's reported they sometimes start eating their owner after they've died.

Many transhumanists want to change all this. They want to rid their worlds of biology. They favor concrete, steel, and code. Where once biological evolution was necessary to create primates and then modern humans, conscious and directed evolution has replaced it. Planet Earth doesn't need iniquitous natural selection. It needs premeditated moral algorithms conceived by logic that do the most good for the largest number of people. This is something that an AI will probably be better at than humans in less than two decade's time.

Ironically, fighting the makings of utopia is a coup a half century in the making. Starting with the good-intentioned people at Greenpeace in the 1970s but overtaken recently with enviro-socialists who often seem to want to control every aspect of our lives, environmentalism has taken over political and philosophical discourse and direction at the most powerful levels of society. Green believers want to make you think humans are destroying our only home, Planet Earth—and that this terrible action of ours is the most important issue of our time. They have sounded a call to "save the earth" by trying to stomp out capitalism and dramatically downsizing our carbon footprint.

The most important issue of our time is actually the evolution of technology, and environmentalists are mistaken in thinking the Earth is our only or permanent home. Before the century is out, our home for much intelligent life will likely be the microprocessor. We will merge with machines and explore both the virtual and physical universe as sentient robots. That's the obvious destiny of our species and the coming AI age, popularized by past and present thinkers like Stephen Hawking, Ray Kurzweil, and Homo Deus author Yuval Noah Harari. Hundred million dollar companies in California led by billionaires like Elon Musk are already working on technology to directly to connect our brains in real time to the internet. We may soon not need the planet at all, just servers and a power source like solar or fusion.

Even if we somehow don't merge with machines (because scared governments outlaw it, for example), we will still use the microprocessor and its data crunching capabilities to change our genetic make-up so dramatically, that it could not be called: natural. We will enter the Star Wars age where we literally change our DNA and biological appearance to become alien and creature-like—to fit whatever environment we need to fit. If this sounds crazy, just consider the Chinese geneticist who last year changed a girl's genes in utero, creating the first alleged designer baby.

Whatever we become—as a former journalist for the National Geographic Channel who has passionately covered many environmental stories—I want to first make it clear what I think humans are doing to the Earth. I do believe we are destroying the environment. I do think we are overpopulated in many cities. I also believe there is high likelihood humans are helping to cause climate change. And while I do think we should not needlessly destroy the planet (especially wildlife) or live in man-made polluted wastelands, the last thing we need to do is put the brakes on consumption, procreation, and progress.

What we're doing to the planet is not as important as what we are achieving as a species in the nearing of transition to the transhumanist age. We will save and improve far more lives in the future via bioengineering, geoengineering, and coming technology than damaged ecosystems across the planet will harm. Salvation is in science and progress, not sustainability or preserving the Earth. To argue or do otherwise is to be sadistic and act immorally against humanity's well-being.

Besides, the envisioned transhumanist future is not just a place where humans can live without the constant threats and hostility of a biological world, it's an age where sentient beings can finally overcome pain and misery. Beyond shedding our terminal flesh and living indefinitely, a secondary goal of the transhuman movement is to overcome all or the majority of suffering—both for ourselves and other nonhuman animals. This is why some believe transhumanism—even if it's made up of post-earthers—is the most humanitarian movement out there.

The tools transhumanists use—science, technology, and reason—to accomplish its watershed aims rely on thriving economies, free markets, and innovation. These mostly come from competitive countries trying to become powerful and make money—a lot of it. Increased economic output is nearly always responsible for raising the standard of living, something that has been going up a lot in the last 50 years for just about every nation on Earth. But that could change quickly as governments increasingly enforce strict pro-environmental regulation which slows down industry and commerce. When you force companies to operate inefficiently for lofty ideals, it hurts their bottom lines, and that in turn hurts workers and everyday people. It's a well-known fact that when economies slow down, people increasingly lose property, turn to violence, and put having families on hold.

But the media usually doesn't paint environmental policy this way. In fact, the media is responsible for a lot of the misinformation propping up the environmental movement, which is often at odds with transhumanism. A typical news headline reads: Billionaires and Politicians Trying to Protect the Planet. I have to chuckle. Billionaires and politicians usually have power-hungry ambitions. In general, they don't want people to have access to their wealth, power, or pristine environments—because they want it for themselves. That's why they want walls, borders, ownership, and control of it all. How many people without the resources to even afford housing, healthcare, and food will ever take a vacation to protected land, even if the land is public like a national park. How many hundreds of millions of people in inner cities ever go visit "nature"? They don't.

Modern environmentalism is a fabricated deceit of and for the rich and powerful. It's especially prominent in liberal places like New York City and my home town San Francisco. Sadly, environmentalism is often just a terrible tool to wield power over those of lesser means. The amount of minorities that visit US national parks—only 22 percent—compared to whites is totally out of whack.

Despite the imperfections of capitalism, I continue to support it because it remains the best hope for the poor to improve their standard of life—because at least the individually poor can work hard, be smart, and eventually become rich themselves. This rags to riches phenomenon is not something that can happen in socialist or communist environments, where nearly everyone loses (except the corrupt)—and those losses often lead to starvation and eventual civil war.

Enviro-socialists and their green new deals are some of the worst examples of those trying to bring change about to society. These people produce very little—rarely enough to improve society in any meaningful way—and they promise a pristine planet oblivious to the fact the great majority of people will be harmed, not helped, by such economy-killing policies.

However, there is an alternative to this ugly duopoly system we exist in for the masses. Let's harness the capitalists and use our nation's natural resources to end poverty, spread equality, and get humans to the transhuman age where science will make us all healthier and stronger. America has approximately 150 trillion dollars of uninhabited Federal Land not including national parks that we could divide up among its citizens—that's a half million dollars in net worth of resources to every American—all 325 million of them. As a nation, let's sell this federal land or preferably lease it to the capitalists and corporations who can pay us something in return—a permanent universal basic income, for example. Some call this a Federal Land Dividend. Leased properly, our Federal Land could provide over $1500 a month to every US citizen, giving a household of four $75,000 a year indefinitely.

Naysayers will say the capitalists will forever destroy the land and resources. But over a quarter century, this is unlikely, since all the new capital and innovation from divesting the land will push us far more quickly in the nanotechnology era—an age where we can recreate environments as we please, including those that are destroyed. If you think making plastic with oil is nifty, just wait till we create whole mature forests and jungles in a week's time with coming genetic editing techniques. Also, we'll be able to regrow any animal or plant—including extinct ones—in mass in a laboratory, something that is already on the verge of happening.

But why create the same nature that is so quintessentially cruel, especially as we become transhumans, with perfectly functioning ageless bionic organs and implants in our brains connecting us to the cloud. Let's us create new environments that fit our modern needs. These will be virtual, synthetic, and machines worlds. These new worlds will be far more moral and humanitarian than that of nature. They will be like our homes, cars, and apartments, where everything in it is inanimate or no longer living, and that's why we find sanctuary and comfort in it. If you doubt this, spend a night in the jungle or forest without any comforts or amenities, and see if you survive.

I don't believe in evil, per se, but if there was such a thing, it would be nature—a monster of arbitrary living entities consuming and devouring each other simply to survive. No omnipotent person would ever have the hate in them to create a system where everything wants and needs to sting, eat, and outdo everything else just to live. And yet, that's essentially what the environment is to all living entities. Environmentalists want you to believe nature is sacred and a perfect balance of living things thriving off one another. Nonsense—it's a world war of all life fighting agony and loss—of fight or flight, of death today or death tomorrow for you and your offspring.

It's time to use science and technology to create something better than an environment of biological nature. This begins with admitting the green do-gooder environmentalists are philosophically wrong— and the coming transhumanist age will usher in a world with far less suffering, death, and destruction, even if we have to harm the planet to first get there. Humans must cast off nature, and then they will finally be free of its ubiquitous hostility, misery, and fatalism. Let's rise above the cultural push of environmentalism, because it's antithetical to our future.

******

## 40) Will Licensing Parents Save Children's Lives?

A few years ago, I was at a doctor party, the kind where tired residents drop by in their scrubs, everyone drinks red wine, and discussion centers around medical industry gripes. I wandered over to a group of obstetricians and listened in. One tall blonde woman said something that caught my attention: with 10,000 kids dying everyday around the world from starvation, you'd think we'd put birth control in the water.

The philosophical conundrum of controlling human procreation rests mostly on whether all human beings are actually responsible enough to be good parents and can provide properly for their offspring. Clearly, untold numbers of children -- for example, those millions that are slaves in the illegal human trafficking industry -- are born to unfit parents.

In an attempt to solve this problem and give hundreds of millions of future kids a better life, I cautiously endorse the idea of licensing parents, a process that would be little different than getting a driver's license. Parents who pass a series of basic tests qualify and get the green light to get pregnant and raise children. Those applicants who are deemed unworthy -- perhaps because they are homeless, or have hard drug problems, or are violent criminals, or have no resources to raise a child properly and keep it from going hungry -- would not be allowed until they could demonstrate they were suitable parents.

Transhumanist Hank Pellissier, founder of the Brighter Brains Institute, also supports the idea, insisting on humanitarian grounds that it would bring a measured sense of responsibility to raising kids. In an essay, he notes professor and bioethics pioneer Joseph Fletcher saying that "many births are accidental". Accidentally getting pregnant often leaves women unable to pursue their careers and lives as they might've hoped for and wanted.

Naturally, some environmentalists, such as American educator Paul L Ehrlich, author of landmark book *The Population Bomb*, also advocate for government intervention to control human population, which would be one sure way to help the planet's fragile and depleted ecosystems.

One of the most comprehensive works about the idea of restricting breeding is Peg Tittle's book *Should Parents be Licensed? Debating the Issues*. It's a balanced collection of essays by experts with various views on the subject.

There's no question that some of the ideas of licensing parents make sense. After all, we don't allow people to drive cars on crack cocaine. Why would we allow them to procreate if they want while on it? The goal with licensing parents is not so much to restrict freedoms, but to guarantee the maximum resources to those children that exist and will exist in the future.

Of course, the problem is always in the details. How could society monitor such a licensing process? Would governments force abortion upon mothers if they were found to be pregnant without permission? These things seem unimaginable in most societies around the world. Besides, who wants the government handling human breeding when it can't do basic things like balance its own budgets and stay out of wars? Perhaps a nonprofit entity like the World Health Organization might be able to step in and offer more confidence. I also like the ideas of local communities stepping in to facilitate this idea amongst themselves. Additionally, to help all parents get a license who want one, I would advocate for government programs to assist them to pass the test.

The sad fact is that many children born into poverty end up costing taxpayers billions. Despite all this money spent, a high percentage of those same kids will end up on the streets, in gangs, or in prison after they become adults. With that mind, just as legalization of abortion has helped drive down crime rates, licensing parents would likely have the same effect.

The approximate 10,000 starving child deaths a day that that the aforementioned doctor cited come from various reports and studies, all of which point to the fact that well over 50 million kids have died due to hunger and malnutrition in the last 30 years. That's a lot of kids.

What's more, 15 percent of kids in the US -- the supposed wealthiest country in the world -- suffer from hunger. A large portion of them are born to families that don't have the resources to properly raise a child. After all, if you can't feed a child, you probably shouldn't have one. Licensing would've restricted many of those births until the parents were more able to deal with the challenges of procreation, which is undoubtedly the most intense and serious long term responsibility most human beings will face in their lives.

As a liberty-loving person, I have always eschewed giving up any freedoms. However, in some cases, the statistics are so overwhelming, that at the very least, given the coming era of indefinite lifespans and transhumanist technology, we must remain open-minded to consider how best to move the species forward to produce the happiest and healthiest children for the planet.

Anything less will leave us with millions more preventable deaths and incalculable suffering of innocent kids.

********

# CHAPTER VII: A POLITICIAN THAT WANTS TO LIVE FOREVER

## 41) Why a Presidential Candidate Is Driving a Giant Coffin Called the Immortality Bus Across America

Since the beginning of recording history, it's always been assumed humans could not escape death and the grave. Yet in the 21st Century, millions of people around the planet are coming to the startling revelation that modern science and technology may soon be able to overcome human mortality. This has not escaped the media, which is increasingly reporting on this phenomenon.

Radical longevity science is sound. Recent studies in mice have shown aging is something that can be slowed down, stopped, and most likely even reversed. Longevity activists and transhumanists — those that want to use science and technology to radically improve the human being — are very excited about this and hope to have the choice to live indefinitely in the near future. In fact, as the Transhumanist Party's 2016 US Presidential candidate, I'm basing much of my campaign on spreading the word that aging should be treated like a disease — and not destiny. And I hope to effectively spread this message by driving a 40-foot bus that looks like a giant coffin across America — a stark reminder that aging and death affects us all unless we do something about it.

On my tour called the "Immortality Bus," I'm hoping to share with others that we should support a society and culture that is strongly pro-science and pro-longevity. My team and I plan to have embedded journalists aboard the bus, documenting our trip and enlivening the conversation.

Unfortunately, many people in America and around the world — especially those who believe in afterlives — are neutral or even oppose stopping biological death and aging with science. They feel it challenges what is natural in the human species. Transhumanists call these people "deathists," those who believe and accept that death is a desirable fate.

Complicating matters for transhumanists is the fact that the 535-person U.S. Congress, the current U.S. President, and all members of the U.S. Supreme Court are 100 percent religious and believe in afterlives. Ultimately, this means the American government has little policy incentive to stop death or to put national resources behind life extension science to make citizen's lives far longer and healthier. As a result, nearly 7,000 Americans die every day (over 100,000 die daily worldwide), even though the science is literally out there to stop it. Longevity scientists and transhumanists believe this is an incalculable tragedy.

Reuters reports that renowned gerontologist Dr. Aubrey de Grey, chief scientist at SENS Research Foundation and the Anti-aging Advisor at the U.S. Transhumanist Party, thinks scientists will be able to control aging in the near future, "I'd say we have a 50/50 chance of bringing aging under what I'd call a decisive level of medical control within the next 25 years or so."

It's not just scientists who are advocating for much longer lives. A plethora of life extension and transhumanist activity has been occurring around the world recently. Google's company Calico was recently formed to fight aging. So was startup Human Longevity, a $70 million dollar company founded by longevity entrepreneur J. Craig Venter, X Prize founder Peter Diamandis, and Dr. Robert Hariri. Even billionaires like Oracle's Larry Ellison and Peter Thiel are putting money into trying to live indefinitely.

I'm hoping my Immortality Bus will become an important symbol in the growing longevity movement around the world. It will be my way of challenging the public's apathetic stance on whether dying is good or not. By engaging people with a provocative, drivable giant coffin, debate is sure to occur across the United States and hopefully around the world. I'm a firm believer that the next great civil rights debate will be on transhumanism: Should we use science and technology to overcome death and become a far stronger species?

The Immortality Bus will begin rolling down American highways in very late August stopping at rallies and events, instigating the kind of conversation America and the world needs to challenge outdated cultural ideals— many which are holding science, technology, and medicine back. Robotic hearts, stem cell technology, designer babies, 3D printed organs, and gene therapy are all controversial, but we should pursue them nonetheless because they represent a chance at improved health for the species.

The biggest concern I expect to address on my bus tour is about overpopulation. It's a valid concern. However, I'm not overly worried about life extension science making the world more crowded. Many people assume just because humans stop dying, the planet will become way more overpopulated. But this is not necessarily the case. What many people don't consider is that technology — according to recent reports at the World Bank — has been raising the standard of living for the entire species over the last 30 years, and will continue to do so. As a rule, the higher the standard of living around the world, the less children people have. In fact, if people knew there were going to live far longer than a standard 75-year lifespan — let's say maybe 500 years — many probably wouldn't have children at all for the first few hundred years. Additionally, new green technologies on the horizon will allow us to better distribute the Earth's precious resources to make this planet more ecologically pristine again.

Science is on the cusp of achieving something monumental for human beings — the possibility of overcoming death. The Immortality Bus is a striking symbol aimed at bringing the conversation of life extension science into the mainstream. Like the great bus tours of the 1960's that brought a culture of hippie love to America and the West, the Immortality Bus hopes to bring a culture of desiring far longer lifespans via science.

Perhaps the most important stop on my Immortality Bus tour will be at the U.S. Capitol building, where my team and I plan to deliver a Transhumanist and Life Extension Bill of Rights to the US Congress. The bill will advocate for government policy to support indefinite lifespans in our species, as well as the use of synthetic and robotic technology to live healthier and better.

The Immortality Bus is a humanitarian mission. In fact, it's transhumanitarian mission. We believe in the 21st century everybody has a universal right to a happy and indefinite lifespan, regardless of their heritage, age, or income. Our goal is not only to make your life and the lives of your loved ones better, but to do this for all the people on Earth.

\*\*\*\*\*\*\*\*

## 42) We Need a New Government Agency to Oversee the Search for Immortality

Scientists increasingly agree that we're fast approaching a moment in medicine—probably within 30 years—where we won't just be significantly lengthening human lifespans, but probably conquering death too.

If this is the case, then the 150,000 people+ who die every day on Earth is doubly tragic. We may soon look back and mourn these hundreds of millions—our parents, friends, and loved ones—who just missed the time in history of achieving indefinite lifespans instead of ending up in a grave.

What this all means is science is nearing the final leg of the greatest race it's ever been in. Medicine's main goal will no longer be to just improve health, but to attempt to guarantee survival for every person that exists on the planet. Unfortunately, one significant challenge to medicine succeeding in this noble life extension aim comes from the most ironic and unlikely of places: the Food and Drug Administration (FDA).

On average, a new drug takes at least 10 years from creation to arrival in your cabinet in America. Additionally, Matthew Herper at Forbes reports that about $5 billion is spent on average in developing a new drug. New medical devices—especially life saving ones—take upwards of seven years to hit the market. For patients, some who are dying to get the drugs and devices, this may as well be an eternity. Nearly all of this has to do with the FDA and the bureaucratic labyrinth that exists to make sure medicine is safe in America.

Now don't get me wrong, I also want safe medicine. And for the most part, the FDA does that. But sometimes there are more important things than safe medicine, especially if you're suffering from a debilitating and terminal disease. For example, many people believe access to medicine before they die is more important than whether that medicine is safe or not. And with such a long, laborious, and costly medical approval process in the US, many inventors and companies that would like to create new medicine don't do it because of the prohibitive procedure of bearing a product from conception to sale.

It's no wonder start-up companies are opening shop in Europe and China, where clinical trials costs less and regulations in some cases are more lax. The obvious question is: How long can this continue before another nation becomes the pharmaceutical and medical device global powerhouse?

Imagine if you're a company, and you have a new heart disease drug that you want to create. You'd have to have cash on hand for a decade (or know you could get it) before you might—if the FDA approved you through its multiple stages—to make single sale on a drug. Now imagine you do the same process in Eastern Europe or Asia, and you only need half the cash on hand. You'd have a far better chance at actually bringing a life saving drug to market and making sure you company can survive until it does so.

There are a lot of reasons for the FDA's notorious regulatory hurdles. Rather than blame them specifically, though, it's easier to blame the true culprits—the vampires of capitalism: lawyers. They have made it so that a few deaths from a new drug (even one that helps tens of thousands of people live far better and longer) are enough to make it so that drug makers won't develop or carry the drug. Class action lawsuits are a reputation killer and simply too much a financial burden.

Tort reform, which I strongly support as a US presidential candidate, would have a major impact on the medical development field. But more importantly, we must bear in mind the concept of life hours— the concept that human beings have a certain amount of healthy, productive hours of life in them, and those hours should be protected at all cost. If a drug has made it so that a large section of people benefit and live longer, more productively because of it, then it's okay that a few don't and possibly even die. We must remember the greater good for society and ourselves, and measure life by life hours, and not our feelings.

I understand this type of thinking is not politically correct, but being politically correct is what healthy people have the luxury of doing. Those dying—those having their cells eaten by cancer, or their minds ravished by Alzheimer's, or the blood flow in their arteries stopped by blockage—tend to be more interested in what makes them healthy and what is functional for American medicine.

What's functional, given the amount of red tape the FDA and the legal system has cast all over American medical development in America is one of two choices.

We could attempt a total reboot of the FDA. Fire everyone and rehire those who are worthy. Ditch the old rules. Limit attorneys having anything to do with policy creation. Set new mandates that insist upon preserving people's life hours, not the bandage pharmaceutical culture America has as its main source of medicine. Also, make this new entity not just an institution that monitors and approves new drugs, but have it be an entity that facilitates America to systematically cure every disease on planet Earth—something that is possible to achieve in the next 50 years if enough resources are put into it. And lastly—the final nail in the coffin—change the FDA name to something new. Something bolder.

I support these actions against the FDA in theory. But I know it's not realistic. One doesn't just change such an embedded government institution without massive controversy, Congressional will, and Presidential support. So the other alternative makes more sense: Leave the FDA alone, but create a totally new institution— sanctioned by the government—whose mission is to encourage and green-light the speedy development of experimental drugs and medical devices for the public. Make it law that this new institution's drug and medical device trial period could not be more than a third of the FDA's average. The government sanction for such a new institution would give the needed authority for US citizens to trust (to some extent) the treatments they receive, while always understanding that such medicine was experimental.

The new drug and medical device field could then be split between traditional FDA approved medicine, and medicine that was more experimental and less proven—but cutting edge and, most importantly, available. Companies developing medical products would have the choice of which approval they sought. But both would be available to consumers with prescriptions.

Such a system would stop our innovators and scientists from going overseas, and keep jobs in America. It would also keep America in the running to remain the world leader in medicine as we enter the transhumanist age. Most importantly, it would keep America in the race to save lives without jeopardizing its regard for the public's safety. It transfers the responsibility to consumers, which any government should always strive to do. If consumers didn't want experimental or fast tracked medicines, they simply could go for FDA approved products only.

The reality is that every year millions of super sick and terminally ill people would likely be willing to try experimental drugs and medical devices rather than suffer or die. America should lead the way to help these people by creating a bold new institution that fast tracks these possible remedies and cures. It's the humane thing to do.

********

## 43) Living Forever Has Never Been More Popular

In less than a month, I'll mark the two-year anniversary to my presidential campaign for the Transhumanist Party. My run for the White House was never about winning, but spreading the idea that Americans can achieve indefinite lifespans through science and technology—if only the government were to help out and put significant resources into the anti-aging field.

While the US government still hasn't shown much interest in supporting longevity research for its citizens, the life extension movement is dramatically expanding around the world. Two years ago, the idea of speaking to 1000 longevity advocates in the same convention hall was a pipe dream. Most transhumanist conferences could barely get 100 people in the same room.

Last weekend in San Diego, that all changed. Billed as the biggest life extension festival in history, RAAD Fest took place from August 4-7. Over 1000 participants made it to the sold-out event, making it the largest group of transhumanists and longevity activists ever to assemble in one place.

The success of the festival signals the growing trend of the life extension movement. In the last few years, major companies like Google's Calico and Human Longevity Inc. have formed to combat aging. Additionally, billionaires like Peter Thiel and Larry Ellison have funded longevity and anti-aging initiatives. Even Facebook founder Mark Zuckerberg recently called for science to end all disease this century.

Leading gerontologist Dr. Aubrey de Grey, a speaker at the conference, told me, "It's great to see people wanting to live longer. The field of rejuvenation research is improving and it's an exciting time for science."

RAAD Fest was conceived last year by a group of transhumanists and longevity advocates at the Coalition of Radical Life Extension, a nonprofit organization based out of Arizona. On its website, RAAD Fest says it aims to become the "Woodstock" of the longevity community.

James Strole, the director of the Coalition for Radical Life Extension, told me, "Next year RAAD Fest could be two or three times as big— maybe even bigger. And we're looking into having international RAAD Festivals too."

In addition to crowds, a plethora of film crews wandered the Town and Country Hotel and Resort and convention center interviewing participants and speakers. A number of minor celebrities were also in attendance, like Three's Company actress Suzanne Somers and Canadian entrepreneur Peter Nygard.

Some journalists attending the event talked of almost a cult-like feel to the festival, with cheering during the speeches of anti-aging leader Dr. Ron Klatz, BioViva's patient zero Liz Parrish, Alcor CEO Max More, and Life Extension Foundation founder Bill Faloon. I told the journalists that those people that believe in using science and technology to overcome death are incredibly passionate about it. The life extension community is not about money, or fame, or power. It's about not dying—which is perhaps the most significant enterprise ever taken on by humans and science. It's easy to see why some people dedicate their entire lives to the life extension field.

Last Saturday, my speech centered on my presidential campaign, trying to grow the Transhumanist Party, and my #1 policy goal: taking money from the US military and spending it on medical and science research. I also pushed my atheism on the crowd, noting that the US government is unlikely to fund life extension science so long as all 535 members of Congress, eight Supreme Court justices, and our president continue to publicly insist they believe in an afterlife with God.

Some of the best moments of the event took place outside the venue, at the packed luncheons and dinners. I had a chance to catch up with friends from all over the world and learn of new projects and ideas. For example, Singularity University's Jose Cordeiro is working on improving the virtually-unknown cryonics space in Spain. Ben Goertzel is considering if one of his artificial intelligence creations might be able to run for president someday, and at least reasonably debate in the 2020 elections. And Dr. Aubrey de Grey is planning more extensive research after a significant investment was made into SENS Research Foundation, located in Mountain View. SENS specializes in rejuvenation biotechnologies.

My only complaint with the festival was there weren't enough young people there. With tickets priced between $400 and $900, most millennials surely had a hard time attending.

James Strole told me the 2017 RAAD Fest will have more youth provisions and student discounts so those under 40 years of age will attend in far larger numbers. I feel that younger group is very important to involve in transhumanism—even if they don't have to worry so much about aging yet—as they are so prolific on social media and can dramatically help the longevity movement grow.

With 31 countries represented at RAAD Fest, I heard lots of different languages being spoken. And it wasn't just people speaking. Ray Kurzweil gave the event's keynote speech via a robot on the stage, covering the latest ideas he has on the future and how nanobots will someday soon dramatically improve human health.

********

## 44) To Ensure a Future of Transhumanism, Atheists Should Confront the Deathist Culture Religion Has Sown

In the West, atheism is growing. Nearly a billion people around the world are essentially godless. Yet, that burgeoning population faces an important challenge in the near future—the choice whether to support far longer lifespans than humans have ever experienced before. Transhumanism technology could potentially double our lifetimes in the next 20-40 years through radical science like gene editing, bionic organs, and stem cell therapy. Eventually, life extension technology like this will probably even wipe out death and aging altogether, damaging one of the most important philosophical tenets formal religion uses to convert people: the promise of being resurrected after you die.

About 85 percent of the world's population believes in life after death, and much of that population is perfectly okay with dying because it gives them an afterlife with their perceived deity or deities—something often referred to as "deathist" culture. In fact, four billion people on Earth—mostly Muslims and Christians—see the overcoming of death through science as potentially blasphemous, a sin involving humans striving to be godlike. Some holy texts say blasphemy is unforgivable and will end in eternal punishment.

So what are atheists to do in a world where science and technology are quickly improving and will almost likely overcome human mortality in the next half century? Will there be a great civil rights debate and clash around the world? Or will the deathist culture change, adapt, or even subside? More importantly, will atheists help lead the charge in confronting religion's love of using human mortality as a tool to grow the church?

First, let's look at some hard facts. Most deaths in the world are caused by aging and disease.

Approximately 150,000 people die every day around the world, causing devastating loss to loved ones and communities. Of course, it should not be overlooked that death also brings massive disruption to family finances and national economies.

On the medical front, the good news is that gerontologists and other researchers have made major gains recently in the fields of life extension, anti-aging research, and longevity science. In 2010, some of the first studies of stopping and reversing aging in mice took place. They were partially successful and proved that 21st Century science and medicine had the goods to overcome most types of deaths from aging. Eventually, we'll also wipe out most diseases. Through modern medicine, the 20th Century saw a massive decrease of deaths from polio, measles, and typhoid, amongst others.

On the heels of some of these longevity and medical triumphs, a number of major commercial ventures have appeared recently, pouring hundreds of millions of dollars into the field of anti-aging and longevity research. Google's Calico, Human Longevity LLC, and Insilico Medicine are just some of them.

Google Ventures' President Bill Maris, who helps direct investments into health and anti-aging companies, recently made headlines by telling Bloomberg, "If you ask me today, is it possible to live to be 500? The answer is yes."

Even smaller projects like the musician Steve Aoki supported Longevity Cookbook with its Indiegogo campaign have recently launched, in an effort to get people to eat better to live longer. All these endeavors add to a growing climate of people and their attitudes willing to accept the transhumanist idea that death is not fate. In fact, in the future, death will likely be seen as a choice someone makes, and not something that happens arbitrarily or accidentally to people.

Despite this positive momentum in the anti-aging science movement, changing cultural deathist trends for 85 percent of the world's population may prove difficult. Humans are a species ingrained in their ways, and getting fundamentally religious people to have an open mind to living far longer periods than before—maybe hundreds of years even—could prove challenging.

Recently, a number of transhumanists, including myself who is a longtime atheist, have attempted to work more closely with governmental, religious, and social groups that have for centuries endorsed the deathist culture. Transhumanists are trying to get those groups to realize we are not necessarily wanting to live forever. As science and reason-minded people, we simply want the choice and creation over our own earthly demise, and we don't want to leave it to cancer, or an automobile accident, or aging, or fate.

Of course, for atheists, the elephant in the room is overpopulation. If everyone lives longer, surely the world will become even more crowded than it is. The good news is that scientists generally believe Earth could handle a far larger human population than we have now, without destroying the planet. But we'd need better methods of resource distribution and laws that ensure equality among people. The key to handling a large population likely rests in new green technology, and using it to fix major environmental problems. Meatless meat is a great example. Much rainforest destruction comes from creating pastures for animal grazing. But we could regrow those forests (which would help the greenhouse and ozone layer problems) by creating meatless meat in laboratories and bypassing the need for livestock. I like this for more reasons than one; 150 million animals are slaughtered every day for our consumption. That's a lot of killing that could be avoided.

In the end, longer lifespans and more control over our biological selves will only make the world a better place, with more permanent institutions, more time with our loved ones, and more stable economies. People, including those who are atheist or religious, will always have the choice to die if they want to, but the specter of death from formal religion will no longer be able to be used as a menacing tool for growing a deathist culture and agenda.

*******

## 45) Now that Humans are Living Longer, College Should be Mandatory

Regardless what state you live in, at least some high school education or its equivalent is required by law in America. These mandates ensure most every kid enters adulthood with the skills to read, write, perform mathematics, and be moderately civilized.

But in an era where scientists believe people born today may live to 150 years of age, are we shortchanging ourselves by not requiring a longer, more rigorous period of education for our youth? Is kindergarten to high school really sufficient for centenarians, or is it time to require all American kids start attending college too?

The history of compulsory education in America goes back almost a century. By 1918, every state required kids to attend at least elementary school. Over the following nine decades, states increased their educational requirements, ensuring youth continue their schooling until at least until 15 years of age, but often until 17 or 18. Whether by traditional high school, charter school, or home schooling, in the 21st century the majority of American youth— almost 80 percent—graduate with a high school-level education.

Most of us take this all for granted because education is so enshrined in American culture and social life. Receiving some form of schooling seems to be the one major topic that citizens— regardless of politics, race, ethnicity, religion, and wealth—agree is positive. A lot of this is due to the fact that education improves one's odds of succeeding in the job market. However, that job market is changing quickly. Last decade, the big news was jobs moving overseas to China. Now it's machines taking many of the jobs in America that are left. Experts predict that by 2025 a third of all jobs will be lost to robots and software.

Politicians are fumbling over themselves trying to find a way to stop this job-loss carnage, one that is a real threat to many US citizens. There's no easy answer to the problem, but one thing is for sure, getting everyone to attend college probably can't hurt. It's a well-known fact that higher education generally makes one far more likely to be employed, get better wages, and land the job and career they want. Studies have shown college grads are happier, healthier, remain married longer, and end up considerably wealthier later in life than those who stopped their educations directly after high school.

In fact, if you look closely, it's hard to find any downsides of receiving a higher education at all. Most people are genuinely in favor of the idea of our country spilling over with spunky, self-confident college grads looking to change the world. The promise of discussing Noam Chomsky, String Theory, and Moore's Law with any twenty-something-year-old you meet on the street seems refreshing.

Of course, college is about much more than just scholastic education. It's also about interacting with professors, debating peers over controversial books, and choosing a worthy major. For many, college also goes hand in hand with wild parties, foreign travel, drugs, new philosophies, and sexual exploration. It's no wonder many people call college the best years of their lives.

So why doesn't society legally mandate such a universally positive experience? Why do we stop at high school and leave the main course of educational development on the table, untouched?

Part of the problem has to do with lifespans. Between the 1920s and 1960s when many states passed the bulk of their compulsory high school education laws, lifespans averaged about 63 years of age. That left an 18-year-old high school grad with paltry 35 years to find a spouse, have babies, make a career, and get prepared for a decade long retirement before dying.

What a difference a few generations make. Most youth today expect to live to at least 100. Marriage is in slow decline. Retirement seems boring. And increasingly men and women are seeing IVF culture as a safeguard to push back having children until their late 30s. All this leaves much more time for pursuits like travel, professional ambitions, education, and even just simple loafing. Extended longevity and advancing reproductive science are wonderful things, but they're really just the tip of the iceberg. In the future, expect these trends to sharply accelerate, giving both women and men much more time in their 20s and 30s to figure out what they want to do in life.

Here lies the real conundrum with upcoming generations. With all this extra free time, doesn't it make sense to require our youth to educate themselves more? It will only help them figure themselves out more and give them the skills to reach whatever dreams they want. Unfortunately, the problems with such a proposal are deep and multifaceted. For starters, such a proposal reeks of authoritarianism. It's understandable to demand 17-year-old go to school. But a 21-year old who is already a bonifide tax paying adult? Compliance with such a law might be impossible. College dropout rates could soar. Our culture has long established that once a kid hits 18 and is out of high school, they are a free agent—a master of the universe.

Another issue critics will have is the complaint that mandating higher education will mean the Mark Zuckerbergs and Tiger Woods of the world won't be able to start their companies or turn pro in their sports, since they'll be forced to go to school. An easy way around that dilemma is to create a college equivalent test, similar to the high school GED. Kids that are smart enough to pass can skip out on school if they don't want to go. They'll miss out on the fun, but at least society will know they've got the smarts to succeed.

The biggest issue about compulsory higher education, though, is affordability: Who is going to pay for almost 20 million American youth to go to college? You won't convince an entire generation of students to take out loans. America is already facing a serious school loan crisis. Maybe, society could incentivize attending college. Might we pay students to get an upper education, like some places in Europe? Or what about offering significant tax benefits, or even creating a monster sister-bill to the existing GI Bill?

Another idea could involve lessening prison operation costs across America. According to a recent report, we spend four times the amount on the American prison system than on education. Many convicts are between 18-22 years of age. Perhaps college would keep them out of prison and significantly drop incarceration and judicial costs. It might be enough to help foot a compulsory college education bill while also improving crime rates around the country.

As challenging as financial considerations for all this might be, the flipside of the coin and its liabilities might be more daunting. In an age where jobs are being lost to machines and China may already be academically our superior, perhaps America needs to dig deeper into its pockets to make its kids smarter. Perhaps the more important question is: In order to protect our future and our nation, can we afford not to have all our youth receive some higher education?

It's worth mentioned too that getting a college education is no longer exclusively living in dorms and learning in brick and mortar classrooms, many of which are halfway across the country. Education is moving online in a big way; virtual classrooms are popping up everywhere. Already, almost all higher education institutions, from small liberal arts colleges to the Ivy League, now offer online classes. Many future students might not even need to leave their homes to get a bachelor's degree. Online education is generally far cheaper than attending a traditional college, and this could significantly help with compulsory upper education costs.

America is entering one of the most challenging times it's ever faced. We are up against increasing wealth inequality, frightening climate issues, and growing technological dominance over nearly every aspect of our lives. Are we going to shortchange our youth because we can't afford it? Or because it's too bossy of us? Or because it requires going against decades of institutionalized culture. Perhaps it's time to ante up: Build new colleges. Hire new teachers. Forge new curriculums. And create a country full of the smartest, brightest, most inquisitive minds on the planet. The brains you insist on our youth having now will carry us all later, no matter how terrifying or beautiful the future becomes. For almost a century America has held that education is a necessity, but we should be cognizant of increasing the length of that education as our citizen's lifespans increase.

*******


## 46) Immortality Bus Delivers Newly Created Transhumanist Bill of Rights to the US Capitol

After months of traveling across the country on a national bus tour, the coffin-shaped Immortality Bus drove into Washington DC and successfully delivered a newly created Transhumanist Bill of Rights to the US Capitol. The delivery of the futurist-themed bill—which aims to push into law cyborg and anti-aging civil rights—ends a national tour for the bus that began its journey in San Francisco on September 5th, 2015.

Crowdfunded on Indiegogo, the provocative Immortality Bus has caught the attention of both America and the world. Highlighted in major media ranging from a 10,000+ word feature in The Verge to a CraveCast podcast on CNET to a leading front page story on BBC.com, the Immortality Bus has been making waves with its controversial message: Science and technology can overcome human death—and will likely do so in the next 25 years.

Much of my US Presidential campaign for the Transhumanist Party—which used the bus as a vehicle to spread a techno-optimistic message—also reiterates this same immortality agenda. I believe if the US Government would dedicate $1 trillion dollars to life extension, longevity, and anti-aging industries, we could likely soon conquer death as a species. For some a trillion dollars may seem like a lot of money, but consider that the US government will spend approximately $6 trillion dollars all together on the Iraq War. Surely, overcoming death through modern medicine and science for all Americans seems a much better idea that fighting far-off wars in places most of us will never visit.

With this in mind, the Transhumanist Bill of Rights seeks to declare that all Americans (and people of all nationalities, as well) in the 21st Century deserve a "universal right" to live indefinitely and eliminate involuntary suffering through science and technology. Those ideas are conveyed in Articles 1, 2 and 6 of the one-page bill. Specifically, Article 6 establishes that we should seek to treat "aging as a disease," something a number of leading gerontologists, like Dr. Aubrey de Grey—Chief Scientist of SENS Research Foundation and the Transhumanist Party Anti-aging advisor—also endorses.

The Transhumanist Bill of Rights—which I read out loud at the steps of the US Capitol, then posted it to the building, and then also hand delivered it to Senator Barbara Boxer's office (my California representative)—covers a number of essential futurist civil rights topics. The bill mandates we protect our species and the planet from existential risk (including environmental destruction, rogue artificial intelligence, and the 25,000 nuclear weapons that currently exist). The bill also calls for renewed commitment to space travel, as well as a government's promise to not put cultural, ethnic, or religious policies before the general health and longevity of its citizens.

Finally, the bill emphasizes the right to morphological freedom: the right to do with one's body whatever one wants so long as it doesn't hurt another person. This is especially important in the gene editing / designer baby age, which has recently been the cause of much discussion in the scientific community. Unfortunately, some of this talk has been disturbingly anti-progress with calls for a moratorium on such technology. I strongly disagree with scientific moratoriums—unless they are directly and obviously harming people today—which is one reason why we need a bill of rights to protect the interests of human health, evolution, and progress.

Below is a copy of the Transhumanist Bill of Rights. While the bill has been carefully considered by myself and other transhumanists, and we hope it will be incorporated someway into law by the United States of America and other governments, the bill is not static and may evolve further as science and technologies evolve. Futurists generally believe no bill of rights, declaration, or constitution should ever remain permanent and unbendable in the transhumanist age— an age where science and technology are dramatically and rapidly changing our lives and our experience of the universe.

TRANSHUMANIST BILL OF RIGHTS

Preamble: Whereas science and technology are now radically changing human beings and may also create future forms of advanced sapient and sentient life, transhumanists establish this TRANSHUMANIST BILL OF RIGHTS to help guide and enact sensible policies in the pursuit of life, liberty, security of person, and happiness.

Article 1. Human beings, sentient artificial intelligences, cyborgs, and other advanced sapient life forms are entitled to universal rights of ending involuntary suffering, making personhood improvements, and achieving an indefinite lifespan via science and technology.

Article 2. Under penalty of law, no cultural, ethnic, or religious perspectives influencing government policy can impede life extension science, the health of the public, or the possible maximum amount of life hours citizens possess.

Article 3. Human beings, sentient artificial intelligences, cyborgs, and other advanced sapient life forms agree to uphold morphological freedom—the right to do with one's physical attributes or intelligence (dead, alive, conscious, or unconscious) whatever one wants so long as it doesn't hurt anyone else.

Article 4. Human beings, sentient artificial intelligences, cyborgs, and other advanced sapient life forms will take every reasonable precaution to prevent existential risk, including those of rogue artificial intelligence, asteroids, plagues, weapons of mass destruction, bioterrorism, war, and global warming, among others.

Article 5. All nations and their governments will take all reasonable measures to embrace and fund space travel, not only for the spirit of adventure and to gain knowledge by exploring the universe, but as an ultimate safeguard to its citizens and transhumanity should planet Earth become uninhabitable or be destroyed.

Article 6. Involuntary aging shall be classified as a disease. All nations and their governments will actively seek to dramatically extend the lives and improve the health of its citizens by offering them scientific and medical technologies to overcome involuntary aging.

*******

## 47) How Brain Implants (and Other Technology) Could Make the Death Penalty Obsolete

The death penalty is one of America's most contentious issues. Critics complain that capital punishment is inhumane, pointing out how some executions have failed to quickly kill criminals (and instead tortured them). Supporters of the death penalty fire back saying capital punishment deters violent crime in society and serves justice to wronged victims. Complicating the matter is that political, ethnic, and religious lines don't easily distinguish death penalty advocates from its critics. In fact, only 31 states even allow capital punishment, so America is largely divided on the issue.

Regardless of the debate—which shows no signs of easing as we head into elections—I think technology will change the entire conversation in the next 10 to 20 years, rendering many of the most potent issues obsolete.

For example, it's likely we will have cranial implants in two decades time that will be able to send signals to our brains that manipulate our behaviors. Those implants will be able to control out-of-control tempers and violent actions—and maybe even unsavory thoughts. This type of tech raises the obvious question: Instead of killing someone who has committed a terrible crime, should we instead alter their brain and the way it functions to make them a better person?

Recently, the commercially available Thync device made headlines for being able to alter our moods. Additionally, nearly a half million people already have implants in their heads, most to overcome deafness, but some to help with Alzheimer's or epilepsy. So the technology to change behavior and alter the brain isn't science fiction. The science, in some ways, is already here—and certainly poised to grow, especially with Obama's $3 billion dollar BRAIN initiative, of which $70 million went to DARPA, partially for cranial implant research.

Some people may complain that implants are too invasive and extreme. But similar outcomes—especially in altering criminal's minds to better fit society's goals—may be accomplished by genetic engineering, nanotechnology, or even super drugs. In fact, many criminals are already given powerful drugs, which make them quite different that they might be without them. After all, some people—including myself—believe much violent crime is a version of mental disease.

With so much scientific possibility on the near-term horizon of changing someone's criminal behavior and attitudes, the real debate society may end up having soon is not whether to execute people, but whether society should advocate for cerebral reconditioning of criminals—in other words, a lobotomy.

Because I want to believe in the good of human beings, and I also think all human existence has some value, I'm on the lookout for ways to preserve life and maximize its usefulness in society.

One other method that could be considered for death row criminals is cryonics. The movie *Minority Report*, which features precogs who can see crime activity in the future, show other ways violent criminals are dealt with: namely a form of suspended animation where criminals dream out their lives. So the concept isn't unheard of. With this in mind, maybe violent criminals even today should legally be given the option for cryonics, to be returned to a living state in the future where the reconditioning of the brain and new preventative technology—such as ubiquitous surveillance—means they could no longer commit violent acts.

Speaking of extreme surveillance—that rapidly growing field of technology also presents near-term alternatives for criminals on death row that might be considered sufficient punishment. We could permanently track and monitor death row criminals. And we could have an ankle brace (or implant) that releases a powerful tranquilizer if violent behavior is reported or attempted.

Surveillance and tracking of criminals would be expensive to monitor, but perhaps in five to 10 years time basic computer recognition programs in charge of drones might be able to do the surveillance affordably. In fact, it might be cheapest just to have a robot follow a violent criminal around all the time, another technology that also should be here in less than a decade's time. Violent criminals could, for example, only travel in driverless cars approved and monitored by local police, and they'd always be accompanied by some drone or robot caretaker.

Regardless, in the future, it's going to be hard to do anything wrong anyway without being caught. Satellites, street cameras, drones, and the public with their smartphone cameras (and in 20 years time their bionic eyes) will capture everything. Simply put, physical crimes will be much harder to commit. And if people knew they were going to be caught, crime would drop noticeably. In fact, I surmise in the future, violent criminals will be caught far more frequently than now, especially if we have some type of trauma alert implant in people—a device that alerts authorities when someone's brain is signaling great trouble or trauma (such as a victim of a mugging).

Inevitably, the future of crime will change because of technology. Therefore, we should also consider changing our views on the death penalty. The rehabilitation of criminals via coming radical technology, as well as my optimism for finding the good in people, has swayed me to gently come out publicly against the death penalty.

Whatever happens, we shouldn't continue to spend billions of dollars of tax payer money to keep so many criminals in jail. The US prison system costs four times the entire public education system in America. To me, this financial fact is one of the greatest ongoing tragedies of American economics and society. We should use science and technology to rehabilitate and make criminals contribute positively to American life—then they may not be criminals anymore, but citizens adding to a brighter future for all of us.

********

## 48) We Must Cut the Military and Transition to a Science-Industrial Complex

Many Americans subscribe to the annoying belief that our nation's military-industrial complex is the surest way to remain the wealthiest and leading superpower in the world. After all, it's worked for the last century, pro-military supporters love to point out.

However, America's dependence on warmongering may soon become a liability that is impossible to maintain. Transhumanism, globalization, and outright replacement of human soldiers with robots are redefining the county's military requirements, and they may eventually render defense budgets far smaller than those now. To compensate and keep America spending approximately 20 percent of the federal budget on defense (as we have for most of the last few years), we'll either have to manufacture wars to use all our newly-made bombs, or find another way to keep the American economy afloat.

It just so happens that there is another way—a method that would satisfy liberals and conservatives alike. Instead of always spending more on our military, we could transition our nation and its economy into a scientific-industrial complex.

There's compelling reason to do this beyond what meets the eye. Transhumanist technology is starting to radically change human life. Many experts expect to be able to stop aging and conquer death for human beings in the next 25 years. Others, like myself, see humans merging with machines and replacing our every organ with bionic ones.

Such a new transhuman society will require many trillions of dollars to satisfy humans ever-growing desire for physical perfection (machine or biological) in the transhumanist age. We could keep our economy humming along for decades because of it.

Whatever happens, something is going to have to give in the future regarding military profiteering. Part of this is because in the past, the military-industrial complex operated off always keeping a few million US military members ready on a moment's notice to travel around the world and fight. But there's almost no scenario where we would need that kind of human-power (and infrastructure to support it) anymore.

Increasingly, small teams of special operation soldiers and uber-high tech are the way America fights its wars. We just don't need massive military bases anymore, nor the thousands of companies to support the constant maintenance of ground troops. Such a reality changes the economics of the military dramatically, and will eventually leave it a fraction of its size in terms of personnel and real estate.

We'll still have the need for technology to fight the wars and conflicts we entangle ourselves in, but it'll be mostly engineers, programmers, and technicians who wear the uniform. The coming military age of automated drones, robot tanks, cyberwarfare, and artificial intelligence just doesn't require that many people. In fact, expect the military not just to shrink, but to mostly disappear into ones and zeroes.

Many people think that the beast of a military-industrial complex—made famous by President Dwight Eisenhower's warning against it in his farewell address—appeared only in the last 50 years. However, others persuasively argue that America has been at war 93 percent of the time since the US Declaration of Independence was signed in 1776—so it's been with us from the beginning.

In liberal California where I live, such facts annoy just about everyone I know—except, of course, those who are shareholders and beneficiaries of the defense industry. Thankfully, despite Congress being led by mostly older white religious men, the younger generation clamors for an improved America—one that can keep its economies running smoothly in a more peaceful way.

This is where the scientific-industrial complex comes in and could satisfy most everyone. And best of all, a society of science requires actual people. Lots of them: nurses, scientists, start-up CEOs, designers, technologists, and even lawyers. The advent of modern medicine to treat virtually every ailment—and the whole anti-aging movement, in general—affects all 318 million Americans. Over half of us suffer from health issues that can be improved but often aren't, for a variety of reasons. For example, the US Census Bureau reports that 40 percent of people over the age of 65 suffer from a disability— and for two thirds of them, it's mobility-related issues. And millions are already racking up the symptoms of heart disease that will kill them. And a younger generation is just waiting to explore bionics, chip implants, and how to upgrade their genes to avoid health problems in the future. All this means we have the fodder to reshape the American economy from a militaristic-based one to a type that thrives off scientific and medical innovation.

Instead of spending American money on sending our soldiers to risk their lives for the whims of war, we could be giving civilians the medicine and healthcare they need to live far better and longer. And living longer has unseen benefits, too. In the future, bonafide transhumans won't have to retire if they don't want to. Their bodies will be ageless and made so strong through technology that work and careers may continue indefinitely—and therefore, so will paying taxes. Transhuman existence is a self-fulfilling economic-boom prophesy for both individual and country.

To help create this new mindset in society, I recently delivered a Transhumanist Bill of Rights to the US Capitol as part of my presidential campaign tour. Article 1 of the bill, among other things, aims to establish that a nation would provide a universal right via science and technology for citizens to live indefinitely if they wanted. This, of course, takes socialized medicine one step further, and doesn't just mandate that the government is interested in your well being, but that it's ultimately interested in your permanent survival.

If a nation was to embrace such a universal right to live indefinitely, it would forever change how a nation looks at the individual lives of its citizens. What would follow is a nation's intense build-out of how to improve the health, longevity, and well being of its people. Additionally, the institutions that are constantly drawing on America, like social security and welfare due to disability, would be less taxing.

Currently, the US Constitution (which I personally think needs a significant rewrite for the 21st century) is overly concerned with protection of national sovereignty—which is one major reason why the military-industrial complex is allowed to grow undeterred. If the US Constitution was endowed with precise wording to also protect an individual's health, well-being, and longevity, then a scientific-industrial complex could rise. This new monster institution would legally be mandated to provide the most modern medicine, technology, and science possible to its people.

Shamefully, the Iraq War will cost the US $6 trillion dollars by the time we're actually done paying all our bills—despite the fact that it's highly questionable whether Iraq was ever even a serious national security issue. However, our country undeniably faces a serious national security issue today—in fact, I'd call it a full blown crisis. Nearly 7,000 Americans will die in the next 24 hours from cancer, heart disease, diabetes, aging, and other issues. And the same amount of people will die tomorrow and the day after.

Overcoming disease and aging in the transhuman age will inevitably occur. The question is not if, but when? The answer lies in how much our nation is willing to spend on scientific and medical research—and how soon. But so long as it continues to spend money on the military instead of citizen's health, human beings will die—which is ironic since it's the military that is supposed to protect us (and not inadvertently sabotage us by swallowing funding for bombs instead of medicine). All we need do as a country is change the direction of our spending, from defense to science. If we can transform America into a scientific-industrial complex, we'll still be able to keep our economy chugging along. Let America's new wars be fought against cancer, diabetes, Alzheimer's, and aging itself. It's a win-win, except for body bag and casket makers.

********

## 49) How New Technology Can Help Stop Mass Shootings

Every time a mass shooting occurs in America, gun owners and supporters are blamed for the tragic violence. This month's Parkland school shooting in Florida was no different. Gun control advocates immediately began calling for stricter gun control laws in America. They won't get far, though: More than 300 million guns are floating around our 50 states, and the number is unlikely to be reduced anytime soon, whether by law or not.

Might there be another way to have so-called "gun control" and still have the freedom to have 300 million or more guns? The answers lie in radical technology. In the recent Las Vegas massacre, a lone shooter from the 32nd floor of a hotel room shot and killed 58 people over approximately 10 minutes at a nearby music festival before police could stop him. Had an armed police drone—or at least a security drone with a smoke bomb or flares—been able to get to the shooter's broken hotel windows from the music venue, dozens of lives might've have been saved. The distance from the shooter to the victims was only about 300 yards. Drones can fly 120 miles per hour and could have likely reached the shooter's room in less than 30 seconds.

Using GPS and custom AI algorithms that detect gunfire bursts and explosion sounds from weapons, a waiting police drone could have found and confronted the shooter in probably less than a minute. It could have exploded a nonlethal smoke bomb in the room of the shooter or directly attacked the shooter himself with weapons.

Drones can also be led to exact locations to stop shooters by specialized smoke alarms that sense gun smoke, sending out GPS coordinates to where the shooting is coming from, or they could be notified through smart building technology built into windows and doors that notify authorities when they've been tampered with or broken. The Las Vegas shooter broke two of his hotel room windows to shoot out of, but no one knew until the shooting began.

Much of this technology is essentially here, and the U.S. military is on the cutting edge with its armada of drones and drone tech. Citizens that desire gun control ought to instead insist the government ask its drone-making partners like Boeing, Lockheed Martin, and other contractors to their products into public protection devices—rather than using them for far-off wars. Even better, private companies like GoPro could help jumpstart the personal security field of drones to help halt terrorism and mass shootings in America.

In the future, all public spaces, schools, and events should consider having terrorism-deterrent drones to protect its citizens—both inside and outdoors. This kind of protection is culturally and institutionally no different than a fire alarm, a fire extinguisher, and the fire department—something that dates back two centuries in America. Even libertarians like myself can get behind it.

Beyond drones, there's gun-surveillance technology. Already, police have the ability to see through walls. Why don't public spaces and schools have special viewing technology that easily registers metal-shaped objects from afar? It's not that hard for a special camera or electromagnetic sensor to detect a hidden metal L-shaped handgun-like object or a long metal rifle in a duffle bag. If a weapon is detected in the camera or sensor, it could send a message to a computer to sound the alarm to alert security and electronically lock doors, whether on a campus or in a place of business or government building. The alarm could also immediately notify authorities and launch that newly acquired security drone to check out and possibly stop an imminent tragedy. Some companies are working on this already, and major casinos in Las Vegas are now experimenting with electromagnetic gun surveillance technology in their lobbies.

People are rightfully distressed at being spied upon and the use of surveillance tech, especially since it's becoming so powerful and pervasive. But a modified thermal camera or electromagnetic sensor need not be able to identify people or their faces. Rather, it could be programmed to only detect weapons and nothing else.

The gun control question in America has been escalating in intensity for decades. It's true that there are about as many gun deaths as driving deaths now, which is unfortunate. I believe something should be done about it—and that something involves gun control. But the control need not come from limiting sales and ownership of guns, but rather from limiting criminals being able to carry out violence in public or private spaces. That can best be done by innovative technology—some of which already exists and much more of which can be quickly developed.

********

# CHAPTER VIII: CLOSING ANTI-AGING & TRANSHUMANISM IDEAS

## 50) The Longevity Peace Prize

In early 2018 I had the opportunity to meet with Lars Heikensten, executive director of the Nobel Foundation. We were at Congreso Futuro, South America's major science conference, where I shared the aims of transhumanism. Namely, to allow people to overcome death and to live indefinitely through the use of radical science and technology.

Talking with Heikensten got me thinking about the benefits of a global-reaching public prize for longevity. The transhumanist movement currently receives little recognition outside of scientific circles; having an international award like the Nobel prizes would help it spread, to the benefit of all society.

Why shouldn't we reward efforts and discoveries in longevity with the same accolades we bestow on people who advance other scientific fields, create important bodies of literature, or fight for world peace?

Our societies already reflect an accelerating desire to stop aging. Think of the advances of knowledge, science, and perspectives we could achieve if we had more time and experience. Think of how equipped we would be to fix problems in the future, if we had already lived through the failings and discoveries of past centuries?

As it turns out, I'm not the only one pondering these questions. Earlier this year, I traveled to the Los Angeles headquarters of the XPRIZE Foundation, which awards prizes for "industry-changing technology that brings us closer to a better, safer, more sustainable world." I had been invited along with about 60 other longevity advocates to help develop a possible prize surrounding longevity. I was ecstatic.

The two-day brainstorming event, hosted by founder Peter Diamandis and opened with a keynote by inventor and futurist Ray

Kurzweil, was like a historic secret gathering filled with noted longevity scientists planning on conquering death. Among other luminaires in attendance was Sergey Young, creator of the $100 million Longevity Vision Fund. During some of the heated sessions, medical doctors and anti-aging researchers argued loudly across the conference room about biomarkers, enzymes, telomerase endings of genes, and how far mitochondria might be manipulated.

Frankly, I felt a little out of my element—I'm not a scientist, but a communicator. My original proposal, designed later with the help of Max More, Natasha Vita-More, James Strole, and Bernadeane, was called the Longevity Peace Prize. It would award a one-time $5 million prize to a person or a group who convinces a government to publicly classify aging as a disease.

Radical longevity—also called life extension or anti-aging science—is in the midst of a massive shift. In the last several months, the amount of investment in various longevity-focused products and endeavors has jumped drastically, from millions to many billions of dollars.

In the past year Google upped its investment by $1.5 billion in anti-aging venture with its Calico Life Sciences. This year Bank of America analysts predicted the longevity industry could be worth "at least $600 billion" by 2025.

Gerontologist Aubrey de Grey says we are just a decade or two away from achieving "robust rejuvenation" milestones that will give us treatments to eliminate many diseases, and start to stop aging. Experiments with rodents have already had success with rejuvenating aged organs. Technologists in Silicon Valley are already taking FDA-approved drugs like metformin to make themselves live longer.

Genetic editing therapy, which can reprogram a gene to not age, is a front runner in anti-aging science. Human experiments are already underway to test the technology. Bionic organs (like artificial hearts) and stem cell therapies are also being developed by dozens of companies designing products to expand healthy lifespans.

In spite of major progress in the scientific community of longevity research—and major resources being invested in the field by leading companies and organizations—many people around the world are unaware that radical life extension research exists as a real thing. Many others are downright skeptical of it.

But the science is already here. And it will inevitably impact how we innovate and solve problems in every major field; from climate science, to politics, to education and health care systems. A major international longevity prize would help societies shift their thinking and become more aware about how deeply the field will alter our future, and how much sooner major shifts will occur than is generally acknowledged.

A major international longevity prize would help societies shift their thinking and become more aware about how deeply the field will alter our future.

Historically speaking, prizes targeted at specific issues have garnered a lot of attention, and led to important advances. They also create a public discourse that garners social accountability in a field by improving visibility and allowing more people to weigh in and speak up and participate more in how funding is allocated via awards. The Field's Medal, awarded to mathematicians under 40 years old, has inspired important changes in math departments around the world. A Pulitzer Prize in journalism often forever changes not only a person's career, but the power of influence of a person's work. Even winning an Oscar at the Academy Awards has helped drive cultural trends and changes.

Few goals of humanity could so dramatically alter the lives of humans as extreme longevity. The field is ripe for a significant award. If people start living to the age of 500—which Bill Maris, former head of Google Ventures says is possible—so many current ideas about human life would change, including ones that impact us personally such as marriage, child-rearing, and retirement.

At the XPRIZE Foundation gathering of futurists earlier this year, my Longevity Peace Prize was the only one of the 16 proposals put forth that was based in longevity activism.

The other proposals targeted scientific achievements such as creating specific human longevity biomarkers, targeting dementia with innovative remediation medicine, and carrying 3D printed or stem cell grown hearts around in cryo-boxes for rushed transplants. As these were all heavily science based, they require teams of medical researchers to complete the missions. After a group vote, my award came in eighth and did not make the top-five cut, now being further reviewed by XPRIZE staff for possible award development.

I applaud my science-minded colleagues in the workshop, and I love the amazing work of the XPRIZE. But for longevity culture and awareness to spread around the world, we will need a prize that target audiences outside of the science and medical communities.

Transformative prizes go way beyond their fields; they inspire everyday people to think differently and learn something new about where the world is heading. And just like the Nobel Peace Prize, anybody should be able to win them for doing something important and beneficial for the world, like getting humans to live far longer than they currently do.

Much of the world thinks of death as natural, and that to interfere with it would be to spite nature and future generations. For most of the world's population, there's also a positive religious implication to death and the opportunity to meet one's maker.

I have spent much of my professional career trying to tell people why living dramatically longer is essential to humanity. I don't want to die ever, and I think it's tragic that individual humans are only here for 80 years or so.

It's been a long and ongoing battle for me to convince the public that people should try to live far longer. Currently, governments don't back projects or fund initiatives that consider aging controversial, much less a disease.

And yet billionaire after billionaire is starting to enter the life extension field and invest in it—everyone from Larry Ellison to Peter Theil to Mark Zuckerberg, who recently donated $3 billion dollars to wipe out all disease by the end of the century.

While this is excellent news for the longevity industry, a society that embraces living super long must have more than the one-percenters backing it. It must have the government and national culture supporting it as well. And it needs to be accessible to people no matter who they are.

A high-profile annual global prize for longevity could be just the answer to engaging more people around the world with the developments in the longevity industry. Death and aging impacts everyone, not just the super wealthy, highly-educated populations. Radical longevity should have the same reach, and be available to everyone.

If I could wave a magic wand—and stay within reasonable financial boundaries—I'd create an annual $1 million prize and award it to the person who has had the most positive impact on the public. This award could be for a scientist, a politician, a celebrity, an artist, a philosopher, or just anyone vehemently committed to anti-aging, who creates a true and felt impact in the longevity domain. Importantly, this award committee should focus on longevity pioneers outside of the direct medical or scientific fields.

My vision is for a peace price, not a medical prize, because living indefinitely is not only about aging, it's also about dealing with our biggest problems, like overpopulation, growing inequality, skyrocketing environmental concerns, and trying to improve living standards for the elderly. Extreme longevity will affect social security burdens, religious beliefs, the health care system, and families who might now regularly have multi-generations living under one roof.

An annual $1 million award would require about a $50 million dollars investment, a small nonprofit foundation to manage the award and its finances, and a good publicity team to organize media attention. There would also need to be a well-planned yearly award ceremony. A $50 million investment may seem like a lot of money, but many one-time donations to universities other non-profit organizations far surpass that.

Isn't our future, and how it is shaped, a matter that concerns all of us? If we have something like a major longevity award, people can participate in how we adapt to longevity, rather than deal with decisions made in closed meetings and medical labs. If we are to celebrate the work of those who are trying to get humans to live dramatically longer, we need to be a part of the conversation from the start.

I'm still hoping some wealthy patron, organization, university, or government might establish a yearly longevity prize that the world will cheer on. As the future comes closer, people should be aware of what longevity scientists and advocates are working on, and how the specter of death might one day be significantly diminished—whether you like it or not. We all need to be ready for that reality when it arrives.

******** 

## 51) Second Coming 2.0: Church Taxes Will Help Resurrect Jesus with 3D Bioprinting

China, Russia, and other nations are striving to dethrone America from being the leading nation in science, technology and medicine. Americans need a better way to finance their research and breakthroughs to remain the outright global science leader. It's not coming from President Trump and his "Make America Great Again" administration, as he recently tried to cut the budget at the National Institute of Health and other science-oriented government agencies, leaving thousands of researchers disheartened.

However, there is a method that could quickly and dramatically boost America's science and technology budget, without further tapping individual taxpayers. Last year, America's 300,000 churches were spared nearly 100 billion dollars in owed taxes through a century-old IRS loophole that doesn't tax charitable church entities. This annual 100-billion-dollar amount—almost triple that of the National Institute of Health's budget—could transform science and medicine in the United States. In fact, in over a decade's time, it could fundamentally alter the nature of disease and disability in our species. With just 10 years of collected church taxes—about one trillion dollars—targeted properly, US scientists could possibly eradicate most major medical ailments, including aging.

Many scientists and researchers today believe we are on the verge of paradigm shifting technologies and cures, whether it be a vaccine for cancer, genetic editing of DNA to end cellular degeneration, or bionic hearts to overcome the world's leading killer: cardiovascular disease. The transhumanist era, where innovation eliminates most physical and mental hardships, is no longer science fiction—it's almost here. The problem is that if it takes 30 years from today to get there versus 10 years, nearly a billion extra people will suffer from disease and die. That's why many modern researchers feel they are in a literal race to save people—to prevent them from biologically suffering and ultimately perishing.

This quest to save people is very similar to nearly all Abrahamic churches. Most religious institutions primary goal is also to save their flock from suffering and prevent them from metaphysically perishing. American science and religion are bound together by this intrinsic commonality.

That's why a growing number of Christians—transhumanist supporters or not—believe that the grace of God is being manifest through the world's scientific creations and progress. When a researcher discovers radical stem cell therapies that allow a paralyzed person to walk again, the work and wonders of Christ are arguably being done in the modern world. The same can be said of controlling Schizophrenia, Alzheimer's or Dementia with drugs and brain implants; instead of losing cognitive abilities in ways that some fundamental Christians believe are prompted by demonic forces, clarity of mind and one's religious-inspired moral compass are fought for and preserved with modern medicine. Today, even the deaf and blind can see through FDA approved devices like the cochlear implant and robotic eye, which both tap directly into the brain. Many disabilities are going extinct—sometimes through the work, prayer, and determination of religious researchers and scientists.

It's easy to recognize many of the scientific breakthroughs happening today are almost identical to the miracles apocryphally performed by Jesus over two millennia ago. Christians especially believe Christ's divine work carries on when the poor receive the cures and medical help. About half of Americans live in a household dependent on the government for some healthcare and financial assistance, and the majority of these people are believers and churchgoers. The fact is Christians often pray for good health, and they increasingly get it.

The sad and unfortunate truth, however, is that Americans could be fulfilling the Biblical Word of God and helping their fellow citizens so much more. If churches would just pay the same taxes other organizations and businesses across America pay, and the US used that money for science and medical research, churches could dramatically do more for their flock within just a few years through improved nationwide health benefits.

However, it goes further than just overcoming ailments and physical suffering. With one trillion extra tax dollars over a decade's time put into specific longevity research, most instances of death itself could likely be stopped. The life extension and anti-aging fields have just a few billion dollars going into them right now, mostly via small nonprofits and start-ups that are already showing minor success in their experiments and research. But if that amount of money was increased by 25 times, the results would be dramatic. We would soon enter the era where humans no longer age or die by disease.

Some theologians and scholars will consider this the manifestation and fulfillment of the Bible's ultimate predictions of Jesus' destiny. It's also precisely this divinity that Christ sought to teach the world of before he was crucified—a world where humans can take shelter in God's alleged power to be spared from harm and evil.

But it doesn't stop at just overcoming suffering and death for those alive today. All those in the past—the approximately 95 billion humans who have lived and died on Planet Earth through the ages—may also be given a second chance at life by the controversial scientific field of Quantum Archaeology, otherwise known as technological resurrection.

Quantum Archaeology combines 3D Bioprinting with the computational power of super computers. Both fields are going through massive disruptions right now. 3D Bioprinters can already print out living tissue, and some believe within 50 years, it will be able to print out entire human beings. Super computers can already do 200,000 trillion calculations per second, and that's before quantum computing arrives in the next few years, which could dramatically increase computational power. In 15 or 30 years—if the microprocessor keeps improving exponentially as it has for almost a half century and it better utilizes the Cloud—a super computer might do millions of quadrillion calculations per one hundredth of a second.

Quantum Archaeologists believe that the universe is mechanistic, leaving the opportunity to reverse engineer parts of our subatomic physical history of the world, including that of a human being's every thought, action, memory, and physical component. If we have enough computing power, and we continue to make progress in discerning modern physics—like our teleportation techniques already in use and the 2013 Nobel Prize winning discovery of the God Particle—we may be able to within the century reverse engineer and record sections of our universe down to the very quarks and electrons that comprise them. Some people believe the entire human race and every person's sub-atomic lifetime composition and history can be found and stored in a memory bank approximately nine miles squared in size. From there, we just download exact perfect atomic blueprints of people a few hours before they died and 3D Bioprint them out—then revive them back to life as we would an unconscious person.

Some theologians and Christian transhumanists argue Jesus could fulfill the Book of Revelations and his proposed Second Coming through this type of technology. In fact, some people believe its humanity's greatest responsibility and imperative to use Quantum Archaeology to resurrect Jesus himself—called Second Coming 2.0—so that he can teach humanity once again in person and carry out his work, including that of the prophesized End Times. After all, through Quantum Archaeology, everyone that has ever lived can be brought back to life—or not. It's through this transhumanist tech that the end of the world—or the beginning of it, depending on one's beliefs—will occur.

When asked who I'd first bring back using Quantum Archaeology, I answer: Jesus. Whether you believe he's the son of God or not, he's undeniably the most influential and significant person to our species—and I'd love to ask him questions about life, philosophy, and his impact on humanity. But resurrecting people presents radical scenarios and major challenges. Jewish friends of mine have expressed interest in bringing back Adolf Hitler so he can face war crimes and be imprisoned. There's even the possibility of bringing back spiritual entities, if our collective Judeo-Christian history is accurate. We might print out the serpent in the Garden of Eden or the archangel Michael.

Naturally, in the future, most people will use Quantum Archaeology to bring back dead loved ones, friends, and family members—some who may have tragically and prematurely died. Indeed, my father passed away recently from disease, and I'd like to bring him back so he can see his grandkids and be with my mom again.

The quagmire is that such a grand future rests on resources and funding to achieve these Christ-like miracles. Otherwise, we'll will continue to lose loved ones to cancer, heart disease, aging, and other reversible diseases and tragedies. With just one small compromise by the over 300,000 churches across America—paying their fair share of annual taxes like everyone else—humanity can literally be saved. This may be the 'leap of faith' God really intended for believers. It's a small one to make for America and the health of its people.

**********

## 52) The Future of Libertarianism Could be Radically Different

Many societies and social movements operate under a foundational philosophy that often can be summed up in a few words. Most famously, in much of the Western world, is the Golden Rule: Do onto others as you want them to do to you. In libertarianism, the backbone of the political philosophy is the non-aggression principle (NAP). It argues it's immoral for anyone to use force against another person or their property except in cases of self-defense.

A challenge has recently been posed to the non-aggression principle. The thorny question libertarian transhumanists are increasingly asking in the 21st century is: Are so-called natural acts or occurrences immoral if they cause people to suffer? After all, taken to a logical philosophical extreme, cancer, aging, and giant asteroids arbitrarily crashing into the planet are all aggressive, forceful acts that harm the lives of humans.

Traditional libertarians throw these issues aside, citing natural phenomena as unable to be morally forceful. This thinking is supported by most people in Western culture, many of whom are religious and fundamentally believe only God is aware and in total control of the universe. However, transhumanists—many who are secular like myself—don't care about religious metaphysics and whether the universe is moral. (It might be, with or without an almighty God.) What transhumanists really care about are ways for our parents to age less, to make sure our kids don't die from leukemia, and to save the thousands of species that vanish from Earth every year due to rising temperatures and the human-induced forces.

An impasse has developed among philosophers, and questions once thought absurd, now bear the cold bearing of reality. For example, automation, robots, and software may challenge if not obliterate capitalism as we know it before the 21st century is out. Should libertarians stand against it and develop tenets and safeguards to protect their livelihoods? I have argued, yes, a universal basic income of some sort to guarantee a suitable livelihood is in philosophical line with the non-aggression principle.

However, it's more of a stretch to talk about the NAP in terms of healthcare. Nonetheless, the same new rules could apply. Libertarian transhumanists believe aging is a negative force— something that we did not invite into our lives. Given that lifespans already doubled in the 20th century due to medicine and technology, and may double again for the same reasons in the 21st century, do we begin to see aging—and even dying—as an unwanted and so-called immoral force against our very lives?

I believe we do. In fact, in my run for the governor of California as a libertarian, a main policy of mine is to label aging as a disease. The classification takes this universal phenomenon and reduces it to exactly what it is: an aggressive force that I do want in my life.

Knowing my arguments, my libertarian friends have asked if I would use government resources to help fight against aging. As a libertarian, I would prefer the private industry to tackle this problem. However, as an aspiring politician in the real world, I understand that when our government and National Institute of Health (NIH) classifies something as a disease, the entire world notices, and often billions of dollars flows into the research to tackle it. I'm not sure about billions of tax dollars being appropriate, but I'm sure I'd want the government stamp of approval—as the people's stamp of approval—on it, making clear that it's an important issue.

I think support for some government help with the fighting of diseases is warranted, if only to be symbolic in support. In my opinion, and to most transhumanist libertarians, death and aging are enemies of the people and of liberty (perhaps the greatest ones), similar to foreign invaders running up our shores. Therefore, I think government and libertarians have some interest in stepping in to protect life and liberties in this case, as they would against foreign aggression.

I'd also argue some government help for the space industry is also warranted. After all, not being able to get humans off this planet easily poses a major existential risk in the event of a global plague, major asteroid hit, or some other catastrophic event. In this case again, a coordinated minarchist state effort against a foreign enemy threatening life, liberty, and country could be acceptable—and not too far of a stretch for some libertarians.

In the end, I'm glad I'm running for governor in California, as I suspect the majority of libertarians will be hesitant at looking at the non-aggression principle in this way. And California has a way of allowing these strange ideas to get the green light and grow. And why shouldn't it? Anything that harms the human being and its ability to thrive is an affront our very lives and values. In the 21st century, we should rise up and use everything within our means to increase the success of our very lives.


********


## 53) Singularity, Life Extension, or Transhumanism: What Word Should We Use to Discuss the Future?


Singularity. Posthuman. Techno-Optimism, Cyborgism. Humanity+. Immortalist. Machine intelligence. Anti-Aging. Biohacker. Robotopia. Life extension. Transhumanism.

These are all terms thrown around trying to describe a future in which mind uploading, indefinite lifespans, artificial intelligence, and bionic augmentation may (and I think will) help us to become far more than just human. They are words you hear in a MIT robotics laboratory, or on a launch site of SpaceX, or on Reddit's Futurology channel.

This word war is a clash of intellectual ideals. It goes something like this: The singularity people (many at Singularity University) don't like the term transhumanism. Transhumanists don't like posthumanism. Posthumanists don't like cyborgism. And cyborgism advocates don't like the life extension tag. If you arrange the groups in any order, the same enmity occurs. All sides are wary of others, fearing they might lose ground in bringing the future closer in precisely their way.

While there is overlap, each name represents a unique camp of thought, strategy, and possible historical outcome for the people pushing their vision of the future. Whatever wins out will be the buzzword that both the public and history will embrace as we continue to move into a future rife with uncertainty and risk, one where for the first time in history, the human being may no longer be classified as a mammal.

For much of the last 30 years, the battle of the best futurist buzzword was fought in science fiction literature and television. Star Trek popularized borg—which helped give commonly used cyborg its meaning. Various short stories and novels tell tales of posthuman civilizations.

The last 15 years marked a shift toward nonfiction work and following of celebrity scientists. Ray Kurzweil's book *The Singularity Is Near* put the term singularity prominently on the word battle map. Biogerontologist Aubrey de Grey's many public appearances touting medical discoveries to conquer human death did the same for life extension science (also called longevity research or the anti-aging field).

The word transhumanism has also long been in use, pushed by philosophers like Max More, David Pearce, and Nick Bostrom. However, until recently, it remained mostly a cult word, used by smaller futurist associations, tech blogs, and older male academics interested in describing radical technology revolutionizing the human experience. Two years ago, a Google search of the word transhumanism—which literally means beyond human—brought up about 100,000 pages. What a difference a few years makes. Today, the word transhumanism now returns almost 2 million pages on Google. And dozens of large social media groups on Facebook and Google+—consisting of every type of race, age group, sexual orientation, heritage, religion, and nationality—have transhuman in their titles. It's also the term that I'm backing, even though I'm not sure it will win out.

Why did this happen so quickly? As with the evolution of most movements and their names, there were numerous moving parts. Dan Brown's international best-seller novel *The Inferno* introduced millions of people to transhumanism. So have media celebrities as diverse as Joe Rogan, Glenn Beck, and Jason Silva, host of National Geographic's *Brain Games*—all three have discussed transhumanism in their work. Even my own relentless writings of transhumanism at search engine-dominating *Huff Post* and *Vice* have helped. A larger reason probably was that both the public and media were ready for an impactful, straightforward word to describe the general flavor of technological existence sweeping over the human race. In case you haven't noticed, the dead live via saline-cooling suspended animation, the handicapped walk via exoskeleton technology, and the deaf hear via brain microchip implants. The age of frequent, life-altering science is now upon us, and transhumanism is the most functional word to describe it.

Even though the words singularity, cyborg, and life extension generate more hits on Google than transhumanism, they just don't feel right describing an ideal and accurate vision of the future. Few people are willing to call themselves a Singularitarian—someone who advocates for a technological event that involves a helpful superintelligence. And Cyborgism is just weird, since the public isn't ready to be merged with machines yet. Life extension isn't bad, but it's generally limited only to living longer.

Almost by default, transhumanism has become the overwhelming leader of the name rivalry. Around the world, a quickly growing number of people know what transhumanism is and also subscribe to some of it. It has become the go-to futurist term to express how science and technology are upending the human playing field.

*******

## 54)  Mind Uploading Will Replace God

I hear a lot of philosophical complaints suggesting that being alive in a computer as an uploaded version of oneself is quite different than being alive in the physical world. While that is open for debate, one aspect of the issue people often forget about is that the so-called spirit world of Abrahamic faiths—which approximately four billion people follow—is based on something at least as odd as the bits in software code that will make up any virtual existence.

When you think about it, trying to wrap your brain around how digital technology and all its wonders are even possible is simply bizarre. Only a tiny fraction of the world's population understand such things in any depth. And an even smaller amount of people actually know how to design and create the microchips, circuit boards, and software that constitutes this stuff in the real world. Human beings are a species dependent on a tech-imbued lifestyle that none of us really understand, but accept wholeheartedly as we go on endlessly texting, Facebooking, and video conferencing.

As a non-engineer atheist grappling with the implications of 1s and 0s manifesting all digital reality, I have at least this much in common with religious people—because they can't understand the spirit world either, even if they insist it exists.

The major difference between the religious spirit world and the digital world is that engineers—technology's high priests—can recreate software, microchips, and virtual environments again and again. They can also test, view, change, manipulate, and most importantly, improve upon their creations. They can apply the scientific method and be assured that the worlds they built of bits and code exist—as surely as we know the Earth is spinning, even if we can't feel it.

People of the planet's major religions can't do this with their spirit worlds. They can only make leaps of faith, and elaborately describe it to you. One either agrees or disagrees with them. Amazingly, proof is not necessary to them.

Being able to upload our entire minds into a computer is probably just 25-35 years off given Moore's Law and the current trajectory of technology growth and innovation. If we can harness the power of artificial intelligence in the next 15 years, then we might get there quicker, as AI will likely make the research and progress happen far more rapidly. But mind uploading is generally considered possible by experts. After all, humans are just material machines, striving to create other machines that mirror ourselves and desires.

Already, interaction between microchip and the brain are occurring all around the world in the form of cranial implants helping the deaf, blind, and mentally ill. Furthermore, telepathy, accomplished recently by researchers in India and France, is the communication medium of the future. We're just in the infancy of all this, but progress is accelerating. I'm looking forward to having an exact copy of myself online one day, both as a companion and as a form of personal immortality in case my biological self dies.

Atheists may not believe in God, but as Sam Harris' recent bestseller, *Waking Up: A Guide to Spirituality Without Religion* points out, we are still deeply spiritual creatures, searching for answers, trying to do good upon the world, and pondering the mysteries of the universe. All this is very healthy, and not that different than some core hopes of the religious-minded. In fact, the only real difference between religious people and atheists is the fact that religious people insist an all-knowing deity is outside of themselves and controlling the shape of the world. Atheists see no God and believe unconscious universal forces and human will are responsible for the shape of the world.

It's that shape of the world that I care about. It's that shape that affects our lives and gives form to our society, nations, and deeds. For millennia, society has been controlled, guided, and manipulated by religion—often for the worse. As a rule, the more fundamental a particular religion was, the better it steered its populace in the direction the leaders of the religion wanted. I often refer to this steering as baggage culture, pieces of social structure, cultural conditioning, and archaic rules passed on from generation to generation with little philosophical change or growth, despite the fact that society evolves every year.

Eventually, such baggage culture weighs us down so much that society becomes lethargic and hopelessly burdened with nonsense in its many actions. This can be seen in the United State's monopolistic two-party pretend democracy system. It can also be seen in Islam—one of the world's fastest growing religions—whose main sacred text, the Koran, is often seen as being at odds with basic modern day women's rights. Of course, one of the most embarrassing examples of baggage culture I know of is America's Imperial measurement system, which favors obfuscation instead of the better metric system.

So what can we do to eliminate our baggage culture? I'm afraid that little will happen as long as we are exclusively biological. Our instincts for vice, petty behavior, and superstition are too strong. There has certainly been a shift towards moral fortitude, reason, and irreligiosity in many developed countries, but it is slow, very slow. The sad truth is we'll be uploading ourselves into machines long before rationality and agnosticism become truly dominant on Earth. The good news, though, is as people begin uploading themselves, they'll also be hacking and writing improved code for their new digital selves—resulting in "real time evolution" for individuals and the species. It's likely this influx of better code will eliminate lots of things that, historically speaking, religion has attempted to protect people from—namely stupidity and social evil.

Take Andreas Lubitz, the co-pilot who likely intentionally crashed Germanwings Flight 9525 in the Swiss Alps, tragically killing all the people aboard. Lubitz is suspected to have been suffering from depression. In the future, we may all have avatars—perfectly uploaded versions of ourselves existing in the cloud or in chip implants in our brains—and these avatars will help guide us and not allow us to do dumb or terrible things. In the Germanwings plane incident, the avatar would have been able to eliminate the depression in itself, and then could've conveyed that to the other, real life self. In this way, the better suited person would've have been given the task of flying the plane.

This may serve what Abraham Lincoln called the better angels of our nature, which we all have but often don't use. Now, with digital clones participating in our every move, someone trustworthy will always be in our head, advising us of the best path to take. Think of it in terms of a spiritual trainer—or even guru—leading us to be the best we can be.

A good metaphor or comparison of this type of digital assistance will already be happening in the next few years when driverless cars hit the road. In the same way driverless cars will help lessen drunk driving, perfected uploaded avatars will also lead us to be more judicious, moral, and reasonable in our lives.

This is why the future will be far better than it is now. In the coming digital world, we may be perfect, or very close to it. Expect a much more utopian society for whatever social structures end up existing in virtual reality and cyberspace. But also expect the real world to radically improve. Expect the drug user to have their addictions corrected or overcome. Expect the domestic abuser to have their violence and drive for power diminished. Expect the mentally depressed to become happy. And finally, expect the need for religion to disappear as a real-life god—our near perfect moral selves—symbiotically commune with us. In this way, the promising future of atheism and its power will reside in achieving this amazing transhumanist technology. Code, computers, and science will one day replace formal religion and its God, and we will be better as a species for it.

*******

## 55) How I Became a Transhumanist

I have a four-foot-tall robot in my house that plays with my kids. Its name is Jethro.

Both my daughters, aged 5 and 9, are so enamored with Jethro that they have each asked to marry it. For fun, my wife and I put on mock weddings. Despite the robot being mainly for entertainment, its very basic artificial intelligence can perform thousands of functions, including dance and teach karate, which my kids love.

The most important thing Jethro has taught my kids is that it's totally normal to have a walking, talking machine around the house that you can hang out with whenever you want to.

Given my daughters' semi-regular use of smartphones and tablets, I have to wonder how this will affect them in the future. Will they have any fear of technologies like driverless cars? Will they take it for granted that machine intelligences and avatars on computers can be their best friends, or even their bosses?

Will marrying a super-intelligent robot in 20 years be a natural decision? Even though I love technology, I'm not sure how I would feel about having a robot-in-law. But my kids might think nothing of it.

This is my story of transhumanism.

My transhumanism journey began in 2003 when I was reporting a story for National Geographic in Vietnam's demilitarized zone and I almost stepped on a landmine.

I remember my guide roughly shoving me aside and pointing to the metal object half sticking out of the ground in front of me.

I stared at the device that would have completely blown my legs off had my boot tripped the mine. I had just turned 30. The experience left me shaken. And it kept haunting me.

That night as I lay tense and awake in my hotel room, I had the epiphany that has helped define the rest of my life: I decided that the most important thing in my existence was to fight for survival. To put it another way: My goal was to never die.

Because I was not religious, I immediately turned to the thing that gave meaning to my world: science and technology. I took a leap of faith and made a wager that day. I later called this (and even later, dedicated a book to it) "the transhumanist wager."

My idea for an immortality wager came from Pascal's Wager, the famous bet that caught on in the 17th century that loosely argued it was better to believe in God than not to, because you would be granted an afterlife if there was indeed a God. My transhumanist wager was based in my belief that it's better to dedicate our resources to science and technology to overcome death while we're still alive—so we don't ever have to find out whether there is an afterlife or not. It turns out I wasn't alone in my passion to live indefinitely through science. A small social movement, mostly of academics and researchers, were tackling similar issues, starting organizations, and funding research.

Some of them called themselves transhumanists.

Fast-forward 16 years from my landmine incident, and transhumanism has grown way beyond its main mission of just overcoming death with science.

Now the movement is the de facto philosophy (maybe even the religion) of Silicon Valley. It encapsulates numerous futurist fields: singularitarianism, cyborgism, cryonics, genetic editing, robotics, AI, biohacking, and others.

Biohacking in particular has taken off—the practice of physically hacking one's body with science, changing and augmenting our physiology the same way computer hackers would infiltrate a mainframe.

It's pretty obvious why it has emerged as such a big trend: It attracts the youth.

Not surprisingly, worrying about death is something that older people usually do (and, apparently, those younger people who almost step on landmines). Most young people feel invincible. But tell young people they can take brain drugs called nootropics that make them super smart, or give them special eye drops that let them see in the dark, or give them a chip implant that enhances human ability (like the one I have), and a lot of young people will go for it.

In 2016, I ran for the US presidency as the Transhumanist Party nominee. To get support from younger biohackers, my team and I journeyed on the Immortality Bus—my 38-foot coffin-shaped campaign bus—to Grindfest, the major annual biohacking meet-up in Tehachapi, California. In an old dentist's chair in a garage, biohackers injected me with a horse syringe containing a small radio-frequency-identification implant that uses near-field communication technology—the same wireless frequency used in most smartphones. The tiny device—it's about the size of a grain of rice—was placed just under the skin in my hand. With my chip, I could start a car, pay with bitcoin, and open my front door with a lock reader.

Four years later, I still have the implant and use it almost every day. For surfers or joggers like myself, for example, it's great because I don't have to carry keys around.

One thing I do have to navigate is how some religious people view me once they understand I have one. Evangelical Christians have told me that an implant is the "mark of the beast," as in from the Bible's Book of Revelations.

Even though I'm tagged by conspiracy theorists as a potential contender for the Antichrist, I can't think of any negatives in my own experiences to having a chip implant. But as my work in transhumanism has reached from the US Military to the World Bank to many of the world's most well-known universities, my chip implant only exasperates this conspiracy.

While people often want to know what other things I've done to my body, in reality becoming a cyborg is a lot less futuristic and drastic than people think.

For me and for the thousands of people around the world who have implants, it's all about functionality. An implant simply makes our lives easier and more efficient. Mine also sends out pre-written text messages when people's phones come within a few feet of me, which is a fun party trick.

But frankly, a lot of the most transformative technology is still being developed, and if you're healthy like me, there's really not much benefit in doing a lot of biohacking today.

I take nootropics for better brain memory, but there's no conclusive research I know of that it actually works yet. I've done some brainwave therapy, sometimes called direct neurofeedback, or biofeedback, but I didn't see any lasting changes. I fly drones for fun, and of course I also have Jethro, our family robot.

For the most part, members of the disabled community are the ones who are truly benefiting from transhumanist technologies today. If you have an arm shot off in a war, it's cyborg science that gives you a robot arm controlled by your neural system that allows you to grab a beer, play the piano, or shake someone's hand again.

But much more dramatic technology is soon to come. And the hope is that it will be available—and accessible—to everyone.

I asked to be added to a volunteer list for an experiment that will place implants in people's brains that would allow us to communicate telepathically, using AI. (Biohacking trials like this are secretive because they are coming under more intense legal scrutiny.) I'm also looking into getting a facial recognition security system for my home. I might even get a pet dog robot; these have become incredibly sophisticated, have fur softer than the real thing (that doesn't shed all over your couch or trigger allergies) and can even act as security systems.

Beyond that, people are using stem cells to grow new teeth, genetic editing to create designer babies, and exoskeleton technology that will likely allow a human to run on water in the near future.

Most people generally focus on one aspect of transhumanism, like just biohacking, or just AI, or just brainwave-tech devices. But I like to try it all, embrace it all, and support it all. Whatever new transhumanist direction technology takes, I try to take it all in and embrace the innovation.

This multi-faceted approach has worked well in helping me build a bridge connecting the various industries and factions of the transhumanist movement. It's what inspired me to launch presidential and California gubernatorial campaigns on a transhumanist platform. Now I'm embarking on a new campaign in 2020 for US president as a Republican, hoping to get conservatives to become more open-minded about the future.

The amount of money flowing into transhumanist projects is growing into many billions of dollars. The life extension business of transhumanism will be a $600 billion industry by 2025, according to Bank of America. This is no time for transhumanism to break apart into many different divisions, and it's no time to butt heads. We need to unite in our aim to truly change the human being forever.

Transhumanists—it doesn't matter what kind you are—believe they can be more than just human. The word "natural" is not in our vocabulary. There's only what transhumanists can do with the tools of science and technology they create. That is our great calling: to evolve the human being into something better than it is.

Because transhumanism has grown so broadly by now, not all transhumanists agree with me on substantially changing the human being. Some believe we should only use technology to eliminate suffering in our lives. Religious transhumanists believe we should use brain implants and virtual reality to improves our morality and religious behavior. Others tell me politics and transhumanism should never mix, and we must always keep science out of the hands of the government.

We need unity of some significant sort because as we grow at such a fast rate there are a lot of challenges ahead. For example, the conservative Christian Right wants to enact moratoriums against transhumanism. The anarcho-primativists, led by people like the primitivist philosopher and author John Zerzan (who I debated once at Stanford University), want to eliminate much technology and go back to a hunting-gathering lifestyle which they believe is more in tune with Earth's original ecology. And finally, we must be careful that the so-called one percent doesn't take transhumanist technology and leave us all in the dust, by becoming gods themselves with radical tech and not sharing the benefits with humanity.

I personally believe the largest danger of the transhumanist era is the fact that within a few decades, we will have created super-intelligent AI. What if this new entity simply decides it doesn't like humans? If something is more sophisticated, powerful, intelligent, and resilient than humans, we will have a hard time stopping it if it wants to harm or eliminate us.

Whatever happens in the future, we must take greater care than we ever have before as our species enters the transhumanist age. For the first time, we are on the verge of transforming the physical structure of our bodies and our brains. And we are inventing machines that could end up being more intelligent and powerful than we are. This type of change requires that not only governments act together, but also cultures, religions, and humanity as a whole.

In the end, I believe that a lot more people will be on board with transhumanism than admit it. Nearly all of us want to eliminate disease, protect our families from death, and create a better path and purpose for science and technology.

But I also realize that this must be done ever so delicately, so as not to prematurely push our species into crisis with our unbridled arrogance. One day, we humans may look back and revel in how far our species has evolved—into undying mammals, cyborgs, robots, and even pure living data. And the most important part will be to be able to look back and know we didn't destroy ourselves to get there.

*******

# APPENDIX

1) A version of *Death Could be a Curable Disease* first appeared in *Metro (UK)*

2) A version of *Transhumanism and Our Outdated Biology* first appeared in *HuffPost*

3) *Silicon Valley Wants to Upgrade Pascal's Wager: New Ideas like Quantum Archaeology are Trying to Challenge Religion and Even the Permanence of Death* first appeared in this book

4) A version of *A World Future Society Conference Speech: Everyone Faces a Transhumanist Wager* first appeared in *HuffPost*

5) A version of *Forget Trump, Zoltan Istvan Wants to be the Anti-Death President* first appeared in *Wired UK*

6) A version of *To Combat Radical Violence in America, We Need Radical Medicine* first appeared in *Vice*

7) A version of *Why We Need a Transhumanism Movement* first appeared in *HuffPost*

8) A version of *Mind Uploading: If Our Thoughts Live Forever, Do We Too?* first appeared in *Quartz*

9) A version of *Transhumanism Is Booming and Big Business Is Noticing* first appeared in *HuffPost*

10) A version of *Death is Not Destiny: A Glimpse into The Transhumanist Wager* first appeared *Singularity Weblog*

11) A version of *When Does Hindering Life Extension Science Become a Crime?* first appeared in *Psychology Today*

12) A version of *Should Transhumanists Have Children?* first appeared in *HuffPost*

13) A version of *Don't Want to Die? Support a 1 Percent Jethro Knights Life Extension Tax* first appeared in *HuffPost*

14) A version of *A TEDx Talk on Life Extension* first appeared in *HuffPost*

15) A version of *Should a Transhumanist Run for US President?* first appeared in *HuffPost*

16) A version of *Origami Cranes: Who is Responsible for this Child's Death? (Introduction to the World's First Mainstream Media Column on Transhumanism: Psychology Today's: The Transhumanist Philosopher)* first appeared in *Psychology Today*

17) A version of *Which New Technology Will Win the Race to Repair and Replace Our Organs?* first appeared in *Singularity University's Singularity Hub*

18) A version of *The Abortion Debate is Stuck. Are Artificial Wombs the Answer?* first appeared in *The New York Times*

19): A version of *The Technology Transhumanists Want in Their Kids* first appeared in *Vice*

20) A version of *A Brain Implant that Registers Trauma Could Help Prevent Rape, Tragedy, and Crime—So Why Don't We Have it Yet?* first appeared in *HuffPost*

21) A version of *The Era of Artificial Hearts Has Begun* first appeared in *Vice*

22) A version of *The Coming Genetic Editing Age of Humans Won't Be Easy to Stomach* first appeared in *Vice*

23) A version of *Can Cryonics, Cryothanasia, and Transhumanism Be Part of the Euthanasia Debate?* first appeared in *HuffPost*

24) A version of *I Visited a Facility Where Dead People are Frozen so They can be Revived Later* first appeared in *Business Insider*

25) A version of *Cryonics, Special Needs People, and the Coming Transhumanist Future* first appeared in *Psychology Today*

26) A version of *'Dead is Gone Forever:' The Need for Cryonics Policy* first appeared in *Newsweek*

27) A version of *Civil Rights Clash: Transhumanists Prepare to Challenge an Anti-Cryonics Law in Canada* first appeared in *HuffPost*

28) A version of *Should I Have Had My Cat Cryonically Preserved?* first appeared in *Vice*

29) A version of *A Tragedy for a Young Cryonicist* first appeared in *Quartz*

30) A version of *The Transhumanism Movement Aims to Eliminate Existential Risk* for the World first appeared in *HuffPost*

31) A version of *We Must Destroy Nukes Before an Artificial Intelligence Learns to Use Them* first appeared in *Vice*

32) A version of *Genetic Editing Could Cause World War III* first appeared in *Vice*

33) A version of *Four Technologies That Could Let Humans Survive Environmental Disaster* first appeared in *Gizmodo*

34) A version of *Space Exploration will Spur Transhumanism and Mitigate Existential Risk* first appeared in *TechCrunch*

35) A version of *Marriage Won't Make Sense When We Live 1000 Years* first appeared in *Vice*

36) A version of *I Visited a Church that Wants to Conquer Death* first appeared in *Business Insider*

37) A version of *Oligarch Pledges $1 Million Prize to the First Person that Can Live to be 123* first appeared in *Daily Mail*

38) A version of *For Christians, Does Being Pro-Life Lead More Souls to Hell?* first appeared in *HuffPost*

39) A version of *Environmentalists are Wrong: Nature Isn't Sacred and We Should Replace It* first appeared in *The Maven*

40) A version of *Will Licensing Parents Save Children's Lives?* first appeared in *Wired UK*

41) A version of *Why a Presidential Candidate Is Driving a Giant Coffin Called the Immortality Bus Across America* first appeared in *HuffPost*

42) A version of *We Need a New Government Agency to Oversee the Search for Immortality* first appeared in *Vice*

43): A version of *Living Forever Has Never Been More Popular* first appeared in *Vice*

44) A version of *To Ensure a Future of Transhumanism, Atheists Should Confront the Deathist Culture Religion Has Sown* first appeared in *HuffPost*

45) A version of *Now that Humans Are Living Longer, College Should be Mandatory* first appeared in *Vice*

46) A version of *Immortality Bus Delivers Newly Created Transhumanist Bill of Rights to the US Capitol* first appeared in *International Business Times*

47) A version of *How Brain Implants (and Other Technology) Could Make the Death Penalty Obsolete* first appeared in *Vice*

48) A version of *We Must Cut the Military and Transition to a Science-Industrial Complex* first appeared in *Vice*

49) A version of *How New Technology can Help Stop Mass Shootings* first appeared in *The Daily Dot*

50) A version of *The Longevity Peace Prize* first appeared in *Quartz*

51) A version of *Second Coming 2.0: Church Taxes Will Help Resurrect Jesus with 3D Bioprinting* first appeared in *The Maven*

52) A version of *The Future of Libertarianism Could be Radically Different* first appeared in *The Daily Dot*

53) A version of *Singularity, Life Extension, or Transhumanism: What Word Should We Use to Discuss the Future?* first appeared in *Slate*

54) A version of *Mind Uploading Will Replace God* first appeared in *Vice*

55) A version of *How I Became a Transhumanist* first appeared in *Quartz*

*******

# AUTHOR'S BIOGRAPHY

With his popular 2016 US Presidential run as a science candidate, bestselling book *The Transhumanist Wager*, and influential speeches at institutions like the World Bank and World Economic Forum, Zoltan Istvan has spearheaded the transformation of transhumanism into a thriving worldwide phenomenon. He is often cited as a global leader of the radical science movement. Formerly a journalist for National Geographic, Zoltan frequently writes for major media, appears on television, and also consults for organizations like the US Navy, XPRIZE, and government of Dubai. His futurist work, speeches, and promotion of radical science have reached hundreds of millions of people. Award-winning feature documentary *IMMORTALITY OR BUST* on his work is now on Amazon Prime. A recent project is his 7-book box set of writings and essays titled the *Zoltan Istvan Futurist Collection*, a #1 bestseller in Essays on Amazon. Zoltan studied Philosophy at Columbia University and the University of Oxford, and now lives in San Francisco with his physician wife and two daughters. Visit his website at: www.zoltanistvan.com

*******

## ABOUT THE BOOK

After publishing his bestselling novel *The Transhumanist Wager* in 2013, Zoltan Istvan began frequently writing essays about the future. A former journalist with National Geographic, Istvan's essays spanned topics from the Singularity to cyborgism to radical longevity to futurist philosophy. He also wrote about politics as he made a surprisingly popular run for the US Presidency in 2016, touring the country aboard his coffin-shaped Immortality Bus, which *The New York Times Magazine* called "The great sarcophagus of the American highway...a metaphor of life itself." Zoltan's provocative campaign and radical tech-themed articles garnered him the title of the "Science Candidate" by his supporters. Many of his writings—published in *Vice, Quartz, Slate, The Guardian, International Living, Yahoo! News, Gizmodo, TechCruch, Psychology Today, Salon, New Scientist, Business Insider, The Daily Dot, Maven, Cato Institute, The Daily Caller, Metro, International Business Times, Wired UK, IEEE Spectrum, The San Francisco Chronicle, Newsweek,* and *The New York Times*—went viral on the internet, garnishing millions of reads and tens of thousands of comments. His articles—often seen as controversial, provocative, and secular—elevated him to worldwide recognition as one of the de facto leaders of the burgeoning transhumanism movement. Here are many of those watershed essays again, organized, edited, and occasionally readapted by the author in this comprehensive nonfiction work, *The Anti-Deathist: Writings of a Radical Longevity Activist.* Also included are some of Zoltan's new writings, never published before. This book is part of a 7-book box set collection of his essential work, the *Zoltan Istvan Futurist Collection*, focusing on futurism, secularism, life extension, politics, philosophy, transhumanism and his early writings. He partially edited the collection during his studies at the University of Oxford. Enjoy reading about the future according to Zoltan Istvan.

*******

www.ingramcontent.com/pod-product-compliance
Lightning Source LLC
Chambersburg PA
CBHW020203200326
41521CB00005BA/230